Inspection of Large Synchronous Machines

Inspection of Large Synchronous Machines

Checklists, Failure Identification, and Troubleshooting

Isidor Kerszenbaum

Department of Electrical Engineering
California State University, Long Beach

IEEE Power Engineering Society, *Sponsor*

IEEE Press Power Systems Engineering Series
Dr. Paul M. Anderson, *Series Editor*

IEEE
PRESS

The Institute of Electrical and Electronics Engineers, Inc., New York

This book may be purchased at a discount from the publisher when ordered in bulk quantities. For more information contact:

IEEE PRESS Marketing
Attn: Special Sales
P.O. Box 1331
445 Hoes Lane
Piscataway, NJ 08855-1331
Fax: (908) 981-9334

For more information about IEEE PRESS products, visit the IEEE Home Page:
http://www.ieee.org/

Printed in the United States of America

10 9 8 7 6 5 4 3 2 1

ISBN 0-7803-1148-5

IEEE Order Number: PC5623

Library of Congress Cataloging-in-Publication Data

Kerszenbaum, Isidor.
Inspection of large synchronous machines : checklists, failure-identification, and troubleshooting / Isidor Kerszenbaum.
p. cm.
Includes bibliographical references and index.
ISBN 0-7803-1148-5 (cloth)
1. Electric machinery, Synchronous—Inspection. I. Title.
TK2731.K47 1996
621.31′33—dc20 95-51730
CIP

To Jackie,
Livnat, and Yigal

Contents

List of Illustrations

Preface

Many synchronous machines that were designed and manufactured in the last decade of the nineteenth century are still in everyday operation. These are mainly hydrogenerators belonging to long-established utilities. The theory of the synchronous machine was already well advanced when these machines were manufactured. Since then, a profuse amount of theoretical literature has been added, especially with the advent of the computer to facilitate the implementation of numerical analysis techniques. Over the years, the continuous push for higher and higher ratings and operating voltages resulted in the development of more complex mechanical structures, cooling arrangements, and insulation materials. As the complexity of the machine increased, design margins became less forgiving. In addition, the growing dependence of society on electric power and the prohibitive cost of failures put the issue of reliability foremost. Reliability has been the driving force for the continuous improvement of techniques and instrumentation designed to monitor the condition of the machine and for the creation of a wealth of industry standards.

Hand in hand with the development of hardware and written literature, a wealth of expertise has been accumulated on the practical aspects of the operation of these machines. This expertise becomes evident during troubleshooting and inspection activities. In spite of all the instruments and written literature, there is no effective substitute for the expert on the spot to troubleshoot or evaluate the condition of the machine during a visual inspection. The recognition of this expertise is demonstrated by the recent implementation of so-called "expert systems" to diagnose problems in large synchronous machines.

In spite of the universal reliance on visual inspection of large synchronous machines as part of their operation and maintenance, there is no written comprehensive compendium to be found on the subject. Succinct guidelines can be found in numerous standards, technical papers, manufacturers' bulletins, and other publications. This book fills this gap by providing a comprehensive reference on the visual inspection of large synchronous machines. It is based on the large body of accumulated experience that can be found in a myriad of publications, the personal experience of the author, and foremost, on the contribution of many associates.

Having the in-house capability to perform reliable inspection of its generators and rotary condensers allows electric utilities and independent power producers to make informed choices regarding repairs/refurbishment of these large machines. This independent capacity for evaluating the condition of the machines can result in substantial savings.

This book was written with the machine's operator and inspector in mind. Although not designed to provide a step-by-step guide for the troubleshooting of large synchronous machines, it serves as a valuable source of information that can prove to be useful during troubleshooting operations. The topics covered are also cross-referenced to other sources. Many such references are included to facilitate the search. Equations describing the operation of the machine were intentionally left out from most of the discussions. There is a vast amount of theoretical literature available for this purpose. The only theory included in this work consists of descriptions of phenomena affecting the reliability of the machine components. In addition, an appendix is included to review the basic concepts of synchronous machine operation. This appendix provides a measure of understanding of how to utilize machine performance characteristics and their sources, and shows how to perform simple circuit calculations.

After a description of site preparation and inspection tools, the book presents a number of forms adequate for entering findings during the inspection of a large synchronous machine. They are written in such a way that most of the items found in major types of machines can be accommodated. Items found only in salient-pole machines are marked *SP,* and items found only in round-rotor machines are marked *RR*. Subsequently, each of the items is described in Chapters 4 through 7 regarding visual appearance and the essence of the processes involved. Figures are introduced when available. The items covered appear under the main members of the machine; that is, stator, rotor, and excitation. A list of the most common electric and mechanical tests performed in the field is presented in Chapter 8. Each test is referenced to the corresponding ANSI/IEEE standards and other publications.

This book can be useful to the machine-designing engineer and systems operations engineer. It provides a wealth of information obtained in the field about the behavior of these machines, including typical problems and conditions of operation. By serving as a source for descriptions of different types of synchronous machines and machine components, it can also be useful to the student of electrical rotating machinery.

The author's intention is to keep updating the contents of this book from his own and others' experience. Therefore, he would appreciate it if readers would please submit their comments or additions to the publisher for incorporation in future editions.

Acknowledgments

The contents of this book are almost impossible to learn in a class. They are the result of personal experience accumulated over years of work with large electric machinery. Most of all, they are the result of the invaluable long-term contribution of co-workers and associates. The following two individuals, in particular, have been a continuous source of knowledge and inspiration over the years, as well as having been kind enough to review the manuscript and provide numerous suggestions: Mr. Jack Cohon, large machines expert, retired from Southern California Edison, and Mr. Tom Baker, Steam Division Electrical Engineer, Southern California Edison.

The author is highly indebted to Professor Alan Wallace of Oregon State University at Corvallis, Professor Radhe Das of California State University, Long Beach, James Edmonds of MCM Enterprises, Ltd., and Geoff Klempner of Ontario Hydro for painstakingly reviewing the manuscript and making important suggestions that were incorporated into the final manuscript.

The author would also like to thank Mr. Dudley Kay, the members of the editorial department of IEEE Press, reviewers, and all other members of the IEEE Press involved in the publication of this book, for their support in making its publication possible.

Isidor Kerszenbaum
Irvine, California

PART 1
Preparation

Chapter 1 **Site Preparation**
Chapter 1 describes how to minimize the risk of machine contamination and to ensure a safe environment in which to perform the inspection.

Chapter 2 **Inspection Tools**
Chapter 2 discusses tools needed to perform the inspection.

Chapter 3 **Inspection Forms**
Chapter 3 includes *The Synchronous Machine Inspection and Test Report,* comprising ten forms applicable to the inspection of most large synchronous machines.

1

Site Preparation

Site preparation is the first significant action to perform immediately before an inspection carried out on, or in the vicinity of, a large electrical machine. Every inspection of a large machine—scheduled or not, long or short—requires a sensible effort toward site preparation. The goal is to minimize the risks of contaminating the machine with any foreign material or object, as well as to ensure a safe environment in which to perform the inspection. Site preparation should be planned ahead of time, and it should be maintained from the moment the machine is opened for inspection until the time it is sealed and readied for operation. Neglecting to prepare and maintain a proper working environment in and around the machine can result in undue risks to personnel safety and machine integrity (see Figs. 1-1 and 1-2).

1.1 FOREIGN MATERIAL EXCLUSION

Foreign material exclusion (FME), a term originated in the nuclear industry, is the set of procedures geared to minimize the possibility of intrusion into the machine of foreign materials before, during, and after the inspection.

In principle, the definition of foreign material is anything not normally present during the operation of the machine that might adversely affect its constituent components if left there. For instance, although ambient air is not necessarily considered a foreign material, the water content of the air is. Water definitively is an extraneous element that should be kept from condensing on the machine

Fig. 1-1 Wooden cover with door at the entrance to the bore area (both sides). These allow control of access to the machine of personnel and tools.

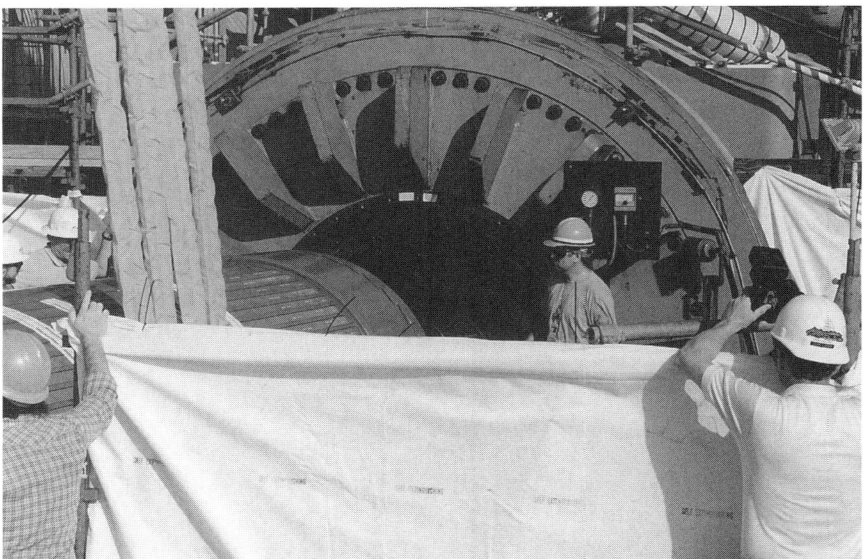

Fig. 1-2 Temporary barrier erected to keep vital components from being contaminated with foreign objects or materials.

windings, retaining rings, and other parts susceptible to mechanical failure from corrosion, or from electric breakdown of the insulation.

Keeping water from condensing onto the machine components can be readily accomplished by containing both stator and rotor under protective covers (i.e., tents), and maintaining a flow of hot air. The hot air and the positive pressure differential inside the tent eliminate the condensation of any significant amount of water (Figs. 1-3 and 1-4). Although the flow of hot air is normally suspended dur-

Fig. 1-3 Typical hot-air blower.

Fig. 1-4 Lowering a protective tent on the rotor of a large turbogenerator.

ing the actual inspection for personnel comfort reasons, thus allowing some condensation to occur, the subsequent flow of hot air will most probably evaporate the moisture and remove it from the containment area [1].

It is important to perform any scheduled electric tests with dry windings. Otherwise, results obtained will not be representative of the winding condition under normal operating conditions.

It is also important not to contaminate the machine inadvertently with corrosive liquids such as solvents, certain oils, and so forth. Sometimes extraneous fluids can be introduced by walking over them and then walking into the machine. Therefore, in situations where stringent FME rules are applied, paper booties are worn over the shoes. Some inspectors prefer the use of rubber booties over their shoes for better and safer grip.

Paper, rubber, or cloth booties will go a long way in eliminating the introduction of small pebbles that may be stuck to the sole of the shoe. When pressure is applied to the end-winding by walking over it, a small pebble can puncture the insulation, thus creating a region of electric-field concentration. This is worth avoiding. It is good practice not to step on the bare coils. A cloth will suffice to protect the winding from the shoe.

The worst enemies of the windings are any foreign metallic objects. They can become airborne due to the high speed of the cooling gas, and break the insulation when striking it. Magnetic particles have been known to cause failures in water-cooled coils by penetrating the insulation over long periods of time, due to the electromagnetic forces acting on the particle. Magnetic as well as nonmagnetic metallic objects may be subject to eddy-current heating, detrimentally affecting the insulation with which they are in contact. Foreign metallic objects such as nails, welding beads, or pens inadvertently left in the bore can short-circuit the laminations of the core. Continued operation under this condition may result in a winding failure due to localized temperature rise of the core. Precautions should be taken to eliminate the possibility of metallic parts or other foreign objects entering the machine. One step in that direction is masking the vent holes of the rotor where these are located outside the stator bore, and covering the rotor when not under inspection or refurbishment (see Figs. 1-5 to 1-7). Metallic objects not required for the inspection should be left outside. This entails removing any coins and other objects (such as medallions, beepers, unnecessary pens, pencils, etc.) from pockets prior to entering the machine. Inspection tools should be carried into the machine on an "as needed" basis. When using mirrors or flashlights in otherwise inaccessible areas, these should be secured by strapping them to one's wrist (Fig. 1-8). In particularly compromising situations, such as with nuclear-powered generators, taking an inventory of tools is recommended both before entering the machine and after exiting it. This is a time-consuming practice, but recommended for all large generator inspections.

Fig. 1-5 Applying masking tape to vent holes of a large 4-pole turbogenerator rotor, with the purpose of eliminating contamination of rotor winding.

Fig. 1-6 Same rotor as previous figure, with the vents covered.

Fig. 1-7 Rotor body covered to avoid contamination while working on other areas of the rotor (end-windings).

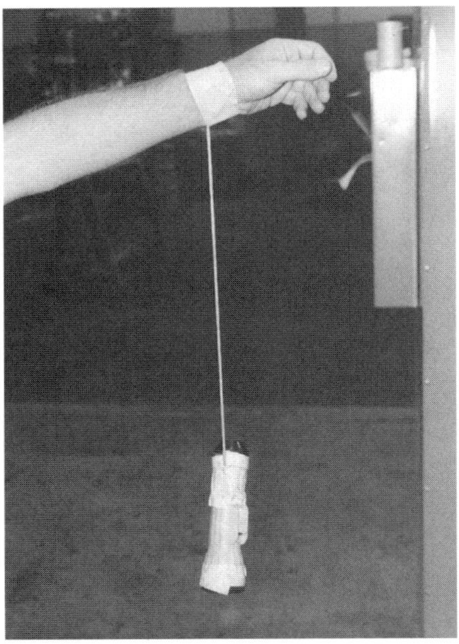

Fig. 1-8 Flashlight strapped to the wrist, to eliminate the possibility of dropping it in inaccessible places.

1.2 SAFETY PROCEDURES—ELECTRICAL CLEARANCES

When carrying out work in an industrial environment, nothing is more important than adhering to all required safety precautions. Large machines opened for inspection often present obstacles in the form of big openings in the floor surface, crevices to crawl through, rods and machine members sticking out, and so on (see Fig. 1-9). They all demand evaluation of required temporary additions to the site, such as beams over the open floor spaces, handrails (Fig. 1-10), secure ladders, and so on.

The obstacles just mentioned are all visible to the people engaged in the inspection. However, an invisible and very powerful element to contend with is the voltage potential (or range of voltages) that may be present in a machine. Although rare, electrical accidents can occur when work is performed in large machines.

A comprehensive inspection of a large machine requires direct physical contact with all windings and other elements that are normally energized during the operation of the machine. "Walking the clearance" is jargon used by some to describe the process of inspecting all breakers, cables, and connections that may be sources of electric power to any part of the machine, and making sure they are deenergized and secure. This means that none of these will be accidentally ener-

Fig. 1-9 Site of a 1350-MVA unit undergoing overhaul.

Fig. 1-10 Provisions allowing safe access to the bore of a large generator.

gized during the inspection. The following is a typical (but by no means all-inclusive) list of safety procedures:

- "Personal grounds" (grounding cables) at both ends of the winding of each phase will minimize the possibility of receiving an unexpected electric discharge (Fig. 1-11).
- Phase leads must be open.
- Neutral transformer (if present) must be disconnected, or have its leads opened.
- Voltage regulators and other excitation equipment must be disconnected.
- Potential transformers are an additional source of voltage to the main windings, and therefore they must be disconnected and secured. Space heaters are often overlooked. To keep the moisture out, space heaters are normally left "on" after disassembling the machine; thus, it is imperative to make sure they are disconnected during the inspection.
- All switches that may energize any part of the machine must be clearly tagged. A tag can *only* be removed by the person who installed the tag originally.
- When inspecting machines with direct gas cooling of the stator windings, discharge resistors are often found on the coil knuckles (see Fig. 1-12).

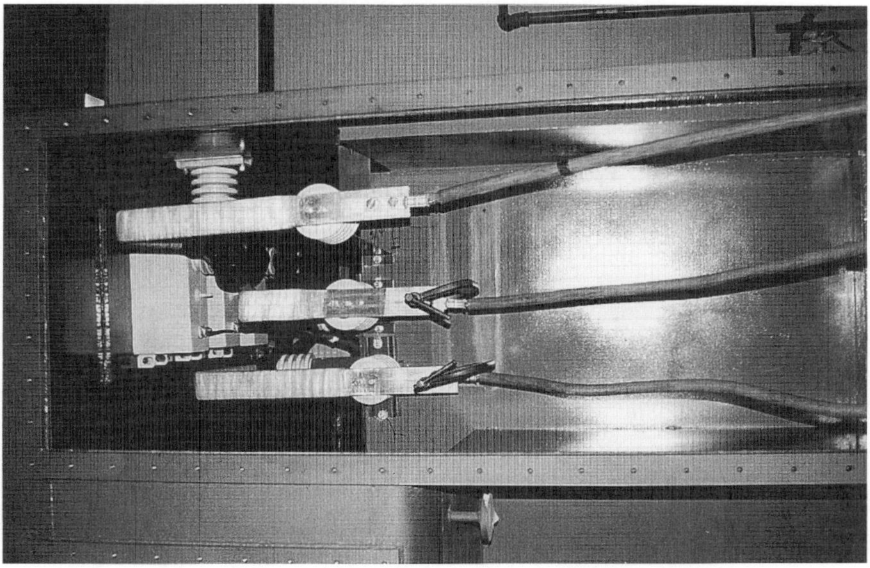

Fig. 1-11 Ground leads applied to a generator unit being overhauled.

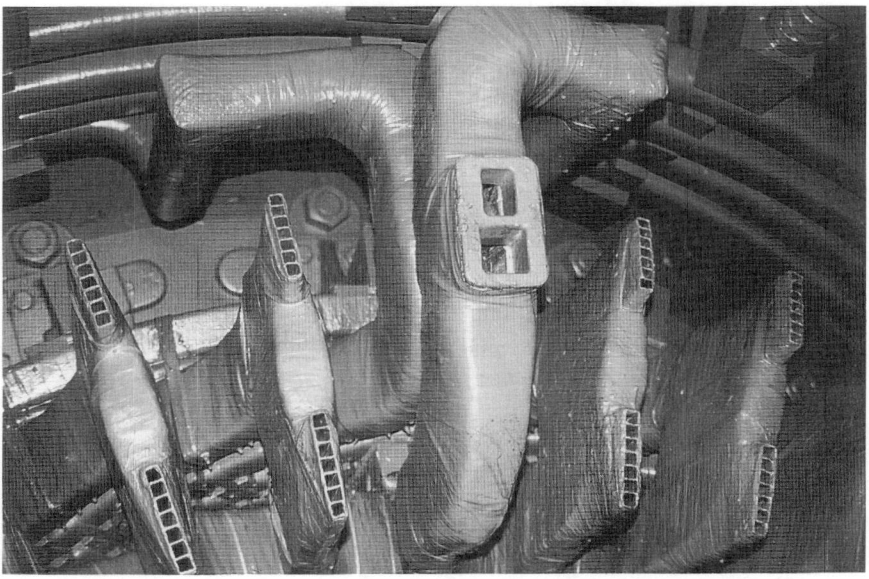

Fig. 1-12 The knuckle area of directly gas-cooled stator coils. The discharge resistors are inside the knuckle area.

When faulty, they may remain charged for substantial lengths of time. Precaution should be taken when inspecting such windings, in particular if high-voltage tests were performed before disassembling the machine. *You can never be too safety-conscious when dealing with high-voltage apparatus.*

- Turning gear must be disconnected (fuses removed) and clearly tagged when inspecting with the rotor in place.
- Gas monitors for confined areas must be used.
- When inspecting hydrogenerators, in addition to the electrical clearance walk, a mechanical clearance walk must be carried out. This should verify the water turbine and valves are locked and secure from inadvertent movement. If possible, the turbine/penstock should be de-watered.
- Additional items as each specific case warrants.
- Follow all relevant safety rules and regulations.

1.3 INSPECTION FREQUENCY

Certain components in large synchronous machines require routine inspections (and sometimes maintenance) between scheduled major outages. Other more comprehensive inspections, requiring various degrees of machine disassembly, are performed during the more lengthy outages.

However, experience has shown that a major inspection after one year of operation is highly recommended for new machines. During the initial period, winding support hardware and some other components experience harsher than normal wear. Retightening of core-compression bolts may also be required during this first outage.

Subsequent outages and inspections can be performed at longer planned intervals. How long an interval? Minor outages/inspections every 30 months, to major outages/inspections every 60 months are typical periods recommended by machine manufacturers. These major outages include removal of the rotor, comprehensive electrical and mechanical (nondestructive) tests, and visual inspections. Obviously, these intervals tend to be longer for machines spending long periods without operation. Most stations have logs containing the actual number of hours the unit was running and the number of starts/stops. This information, together with the manufacturer's recommendations, can be used to schedule the inspections and overhauls.

Large utilities that have many generators in their systems and many years of experience running these machines have formed their own maintenance and inspection criteria and schedules. Although working closely with the respective machines' manufacturers, these utilities tend to extend the periods between outages for those machines that experience has shown to have good records of operation, and shorten the periods between outages/inspections for machines that have been characterized by more frequent failures (as may be the case with old hydrogenerators and rotating condensers).

Frequently, the major outages during which the opportunity presents itself to carry out a major inspection follow the need to maintain the prime mover more than the generator itself.

1.3.1 Condition-Based Maintenance (CBM)

High equipment reliability, high outage costs, and the new competitive outlook permeating the electric power industry have resulted in a new approach to machine maintenance. Condition-based maintenance (CBM) relies heavily on sophisticated on-line instrumentation and evaluation techniques to assess the condition of the machine. In this manner, the periods between major outages/inspections can be increased beyond the fixed, scheduled intervals of the past. The main concept is to base maintenance on the actual condition of the machine rather than on a fixed schedule.

One suggested method of obtaining significant information on the condition of the machine is to retrieve temperature, vibration, PD (partial discharge) activity, air/gas-gap flux, and other readings on the machine under load (prior to shutdown), and compare them with the same test data obtained on previous occasions under similar operating conditions.

REFERENCES

[1] ANSI/IEEE Std 43-1974, "Recommended Practice for Testing Insulation Resistance of Rotating Machinery," Item A2, p. 15.

2

Inspection Tools

Probably the most important item of reference for the inspector is not a gadget or an instrument, but a piece of paper: a record of previous visual inspections and electric tests. Findings from past inspections are like a compass, helping to guide the inspector to areas already proved to be problematic. In his own experience, this author has always felt his work is greatly facilitated when armed with substantial information from previous inspections and tests. Most users of large synchronous machines carry out a minimum set of electrical tests on the machine before disassembling it for a visual inspection. Results of these tests have the double advantage of first, calling attention to problematic areas, and second, allowing comparison with the test results obtained after cleanup and refurbishment performed on the machine. It goes without saying that a comprehensive report on the inspection should always be made and archived. This report will become a very helpful reference during the next inspection, probably several years down the line.

An additional source of information that has served this author well is historical information obtained from identical or similar machines. Still other significant sources of information are the facts/recommendation sheets periodically provided by the manufacturers.

As to the "bag" of tools carried by the inspector, it may include the following (see Fig. 2-1):

- Writing pad attached to a clipboard, and a nonmetallic pen attached to the board with a string.
- Disposable paper or cloth booties, or rubber booties.

Fig. 2-1 Easy-to-carry tool case with the most essential tools for visual inspection of the machine.

- Disposable paper or cloth overall.
- Work shoes with soft sole (rubber).
- Floodlight with an extension cord.
- Flashlights and a set of spare batteries. If a flashlight is at risk of falling to inaccessible places, it should be attached to the wrist with string and tape.
- Clean rags.
- A small sealed container with white rags to be used as swabs; useful for taking samples of contamination when required.
- A set of mirrors with articulated joints and expandable handles. If the mirrors are at risk of falling to inaccessible places, they should be attached to the wrist with a string and tape.
- A hammer with both a soft (rubber) and hard (plastic) heads, for probing wedges, insulation blocking, etc. For a wedge survey of the entire machine, hand-held electromechanical probes are commercially available (see Fig. 2-2). After an initial setting, the probe identifies each wedge as *tight* or *loose/hollow*.
- A set of magnifying glasses or hand-held microscopes to probe for corrosion or electrically originated pitting on retaining rings and in other critical locations.
- Charts from manufacturers of commutator brushes depicting observable signs of bad commutation (Fig. 2-3).

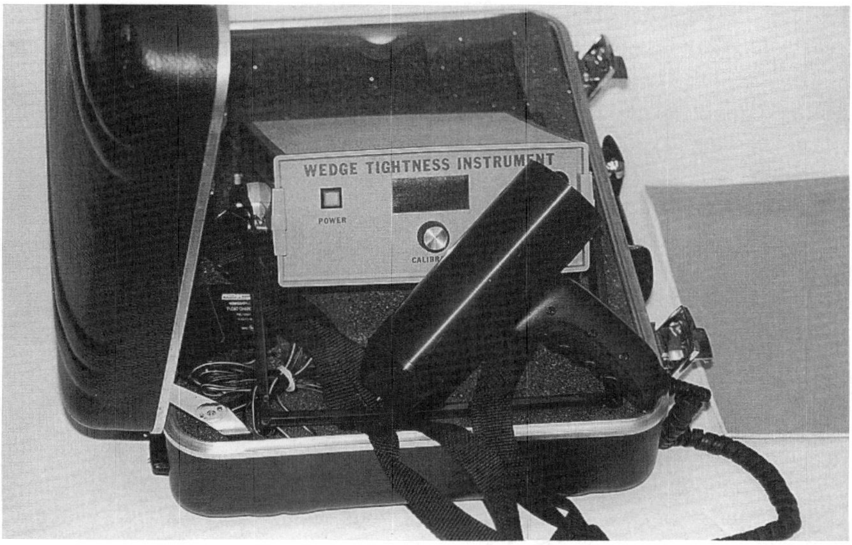

Fig. 2-2 Commercially available wedge tightness electromechanical tester.

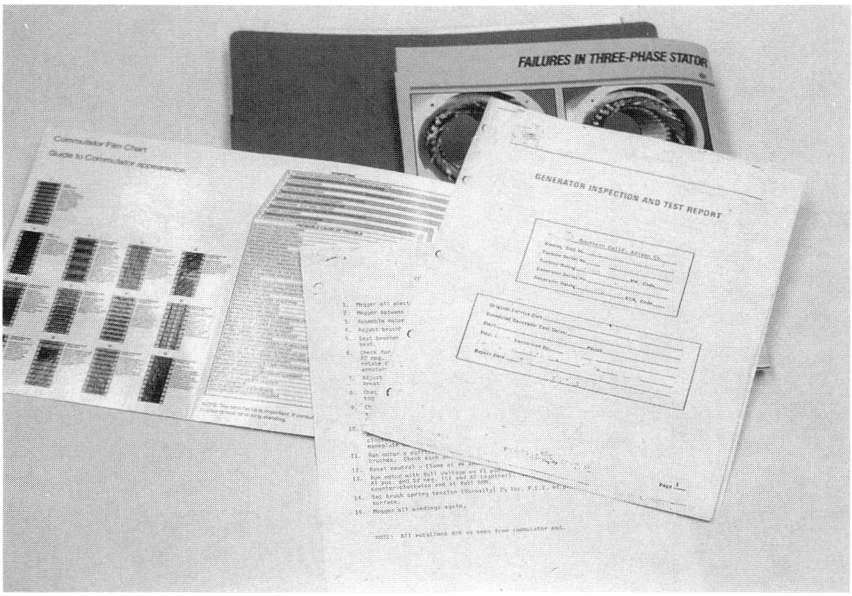

Fig. 2-3 Commutation performance charts from various manufacturers of brushes, and other useful information.

- Boroscope, especially suitable to inspect under the retaining rings, air/gas ducts, air/gas-gap of machines with the rotor in place, and other inaccessible spots (Fig. 2-4).
- A good camera capable of taking close shots of small areas.
- A small magnet for the extraction of loose iron particles.

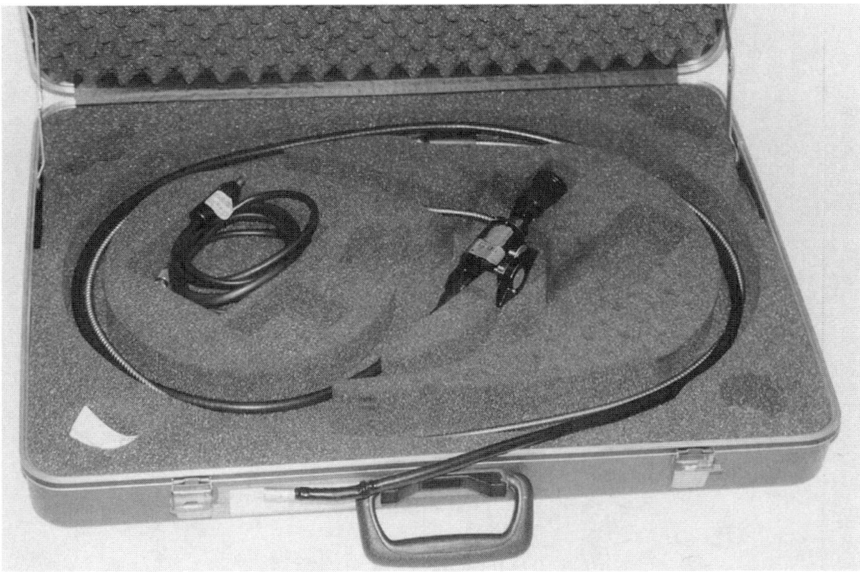

Figure 2-4 Boroscope used for visual inspection of otherwise inaccessible areas of the machine. Light source and optional video equipment not shown.

3

Inspection Forms

Chapter 3 includes the "Synchronous Machine Inspection and Test Report," which comprises ten inspection forms for large synchronous machines. The forms accommodate both salient-pole and round-rotor machines, as well as motors, generators, and rotary condensers. Some individual items on the forms apply to only one of the above; some apply to all of them. Items that are found only in salient-pole machines are marked *SP* on the forms. Items found only in round-rotor (cylindrical) machines are marked *RR* on the forms.

The forms are not intended to cater to every type of machine currently being used in the industry. Their purpose is to serve as an example of how such forms might be organized. However, with minor additions or changes, they can suffice to aid in the inspection of most machines.

Figures 3-1 to 3-3 depict typical synchronous machines encountered across the industry. Figure 3-4 is a schematic representation of a large 2-pole turbine generator. This hybrid-cooled generator has a water-cooled stator and hydrogen-cooled rotor. The bearings are mounted on the machine's brackets (end-shields). Figure 3-5 shows a large 4-pole turbine generator. The machine has stator and rotor windings, both water-cooled. Its bearings are mounted on pedestals. In both Figures 3-4 and 3-5, certain details of the water feed system to the stator windings can be seen.

Figure 3-6 depicts a typical generator and its components. The figure shows the frame construction and the end-winding supports attached to the frame. The figure also shows some of the external details of the cylindrical rotor.

Fig. 3-1 An air-cooled, 35-MVA synchronous condenser.

Fig. 3-2 A hydrogen-cooled, 60-MVA synchronous condenser.

Fig. 3-3 A vertical, air-cooled, synchronous hydrogenerator.

One of the important steps to be taken at the onset of an overhaul is the prioritization of those items to be inspected immediately after disassembly of the machine. For instance, if a wedge survey is considered necessary, it should be carried out as soon as possible after removal of the rotor. The reason for this is the relatively long lead time required to purchase a new set of wedges. Pressure-testing of the hydrogen seals of the rotor of a turbine generator should also be done as early into the overhaul as possible. The same can be said about non-destructive examinations (NDEs) of retaining rings and other critical items that may require replacement.

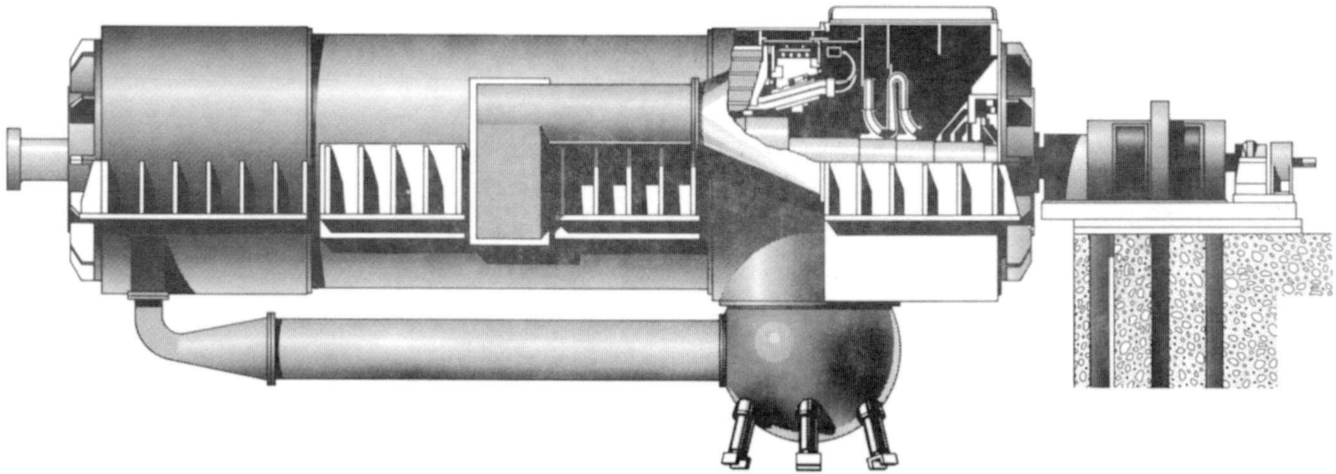

Fig. 3-4 A 1182-MVA, 3000-rpm 2-pole generator (stator winding water-cooled; rotor winding hydrogen-cooled). (Reproduced with permission from "Design and Performance of Large Steam Turbine Generators," 1974, ABB.)

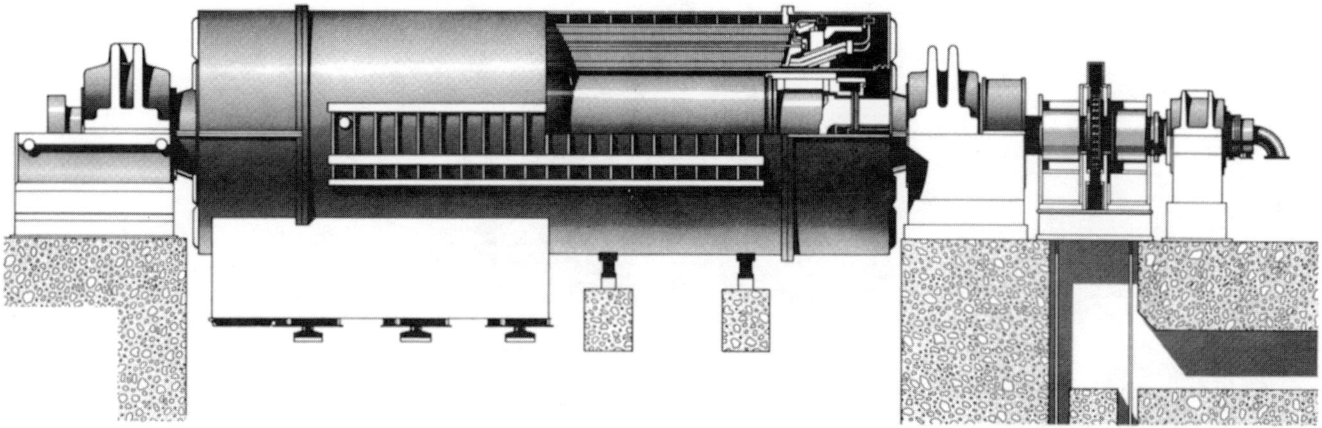

Fig. 3-5 A 1620-MVA, 1500-rpm 4-pole generator (stator and rotor windings water-cooled). (Reproduced with permission from "Design and Performance of Large Steam Turbine Generators," 1974, ABB.)

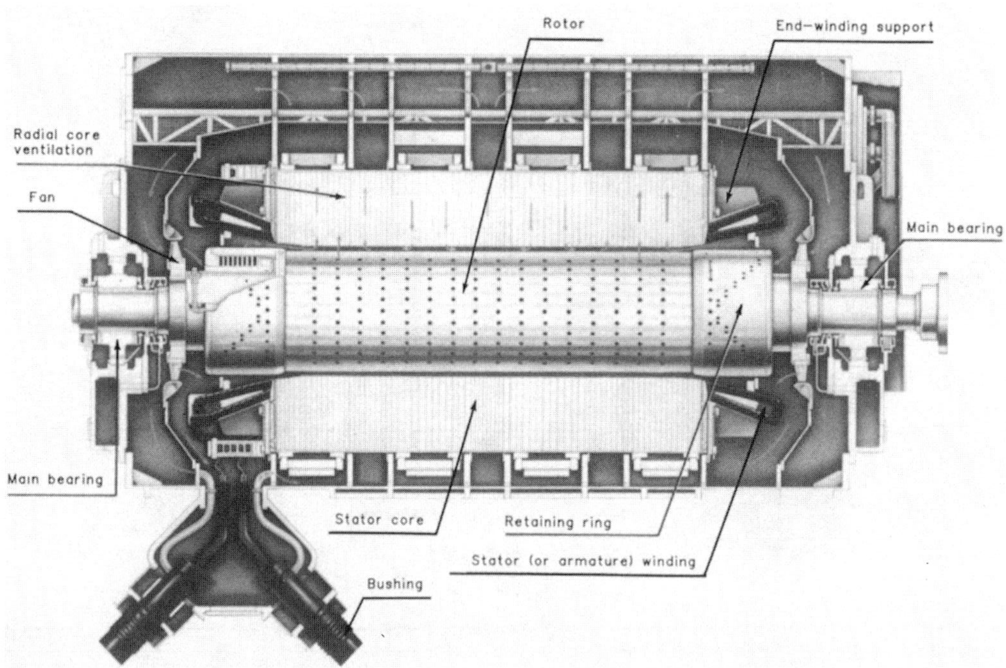

Fig. 3-6 Schematic representation of a typical large turbogenerator. The figure shows details of the frame construction, and the end-winding supports attached to the frame. (Copyright © 1987, Electric Power Research Institute. EPRI EL-5036, *Power Plant Reference Series, Volumes 1–16*. Reprinted with Permission.)

SYNCHRONOUS MACHINE INSPECTION AND TEST REPORT

Form 1: Basic Information

	mark with X
GENERATOR	
CONDENSER	
MOTOR	
SALIENT-POLE	
ROUND-ROTOR	

Station/Company where installed ______________________

Unit no. ______________________

Prime mover type: Steam ____ Gas ____ Hydro ____ Diesel ____

Serial no. of generator ______________________

Manufacturer ______________ Frame ______________

Date of manufacture ____________ Year installed ____________

Date of last rewind: Stator ____________ Rotor ____________

Date of last major inspection ______________________

Total operating hours ______ Operating hours since last overhaul ______

Total number of starts/stops _____ Number of hours in turning gear _____

Present inspection performed by ______________________

Assisted by ______________________

Date of inspection ______________________

Form 2: Nameplate Information

Rated MVA ___ Power factor ___ Short circuit ratio ___ Field I^2t ___

Stator: Line voltage ____________ kV Rated current _________amps

Field: DC voltage___________ volts DC current __________amps

Nominal speed _________rpm No. of poles ___ Frequency _____Hz

Stator cooling: Open air ___ Air/Water ___ Direct water___ Hydrogen ___

Rotor cooling: Air ___ Hydrogen ___ Water ___

Stator insulation: Asphalt ___ Epoxy/Resin-Mica ___ VPI ___ Other _____

Max. H_2 pressure [psi] ____

Form 3: Inspection Accessibility

	Yes	No
Rotor out of bore	☐	☐
Inboard top-bracket removed	☐	☐
Inboard bottom-bracket removed	☐	☐
Outboard top-bracket removed	☐	☐
Outboard bottom-bracket removed	☐	☐
Bushing well open	☐	☐
Inboard retaining ring off	☐	☐
Outboard retaining ring off	☐	☐
Exciter's rotor out of bore	☐	☐
Electrical and mechanical clearance	☐	☐

Additional details: ____

____

____

Form 4: Stator Inspection

Type of blocking: *Maple* ____ ; *Textolite*____ ; *Conforming* ____ ; *Felt* ____

End-windings make-up: *Z-ring* ____ ; *Radius strip* ____ ; *Sausages* ____ ; *Other* ____________

Ties: *Glass roving* ____ ; *Glass cord* ____ ; *Flax* ____ ; *Other* ________________

Type of wedges: *Flat* ____ ; *Piggyback* ____ ; *Other* ________________

Type of side fillers: *Flat* ____ ; *Ripple spring* ____ ; *Nonexisting* ____

Boroscopic inspection of the air-duct area performed: *Yes* ____ ; *No* ____

If wedge survey performed: *% of loose wedges* ____ ; *% of hollow wedges* ____

Number of damaged resistors in nose of water-cooled windings ____

"O" for satisfactory
"X" for unsatisfactory

Item	Description	N/A	O/X	Corrective Action
S01	Cleanliness of bore (oil, dust, etc.)			
S02	Air ducts clogged/unclogged?			
S03	Iron oxide deposits?			
S04	Hardware condition (bolts, nuts, etc.)			
S05	H.V. bushings			
S06	Stand-off insulators			
S07	Bushing vents clogged/unclogged?			
S08	Greasing/red-oxide deposits on core bolts?			
S09	Space heaters			
S10	Fan-baffle support studs			

Stator Inspection (continued)

"O" for satisfactory
"X" for unsatisfactory

Item	Description	N/A	O/X	Corrective Action
S11	Heat exchangers cleanliness			
S12	Heat exchangers leaks?			
S13	Hydrogen desiccant/dryer			
S14	Core-compression ("belly") belts			
S15	Bearing insulation (at pedestal or babbit)			
S16	Coil cleanliness (oil, dust, etc.)			
S17	Blocking condition			
S18	Ties between coils tight?			
S19	Ties between coils too dry?			
S20	Ties to surge-rings tight?			
S21	Ties to surge-rings too dry?			
S22	Surge-rings insulation condition			
S23	Surge-rings support assembly			
S24	Additional end-wdg. support hardware			
S25	RTD and TC wiring hardware			
S26	Asphalt bleeding?—Soft spots?			
S27	Tape separation?—Girth cracking?			
S28	Insulation galling/necking beyond slot?			
S29	Insulation bulging into air ducts?			

Stator Inspection (continued)

"O" for satisfactory
"X" for unsatisfactory

Item	Description	N/A	O/X	Corrective Action
S30	Insulation too dry?—Flaking?			
S31	Circumferential bus insulation			
S32	Corona activity			
S33	Wedges condition (wedge survey below)			
S34	Wedges slipping out at ends?			
S35	Fillers slipping out at ends?			
S36	Bars bottomed in slot?			
S37	Laminations bent in bore?—Broken?			
S38	Laminations bulging into air ducts?			
S39	Terminal box CTs condition			
S40	Bushing-well insulators and H_2 sealant condition			
S41	Winding support bearing-bolt condition (RR)			

RR: Item to be found only in round-rotor machines

Form 5: Rotor Inspection

Type of rotor: *Salient* ____ ; *Round* ____ ;

Retaining rings type: *Magnetic* ____ ; *Nonmagnetic* ____ ; *%Mn* ____ ; *%Cr* ____
Spindle-mounted ____ ; *Body-mounted* ____

Collector rings position: *At same end* ____ ; *At opposite ends* ____

Number of collector rings per polarity: *One* ____ ; *Two* ____ ; *Three* ____

"O" for satisfactory
"X" for unsatisfactory

Item	Description	N/A	O/X	Corrective Action
R01	Rotor cleanliness (oil, dust, iron/copper dust)			
R02	Retaining rings' visual appearance (RR)			
R03	Centering rings' visual appearance (RR)			
R04	Fan-rings' visual appearance			
R05	Fretting/movement at rings' fits?			
R06	Fan blades condition			
R07	Bearing journals condition			
R08	Balance weights/bolts condition			
R09	End-wedges (touching end. rings, loose?) (RR)			
R10	Other wedges (overheated, loose, cracks?) (RR)			
R11	End-windings condition (RR)			
R12	Top series connections (SP)			
R13	Bottom series connections (SP)			

RR: Item to be found only in round-rotor machines
SP: Item to be found only in salient-pole machines

Rotor Inspection (continued)

O/X: "O" for satisfactory, "X" for unsatisfactory

Item	Description	N/A	O/X	Corrective Action
R14	Field-pole keys in dovetail condition, loose? (SP)			
R15	V-shaped inter-pole blocks (SP)			
R16	Insulation between turns (SP)			
R17	Field coils shifted or damaged? (SP)			
R18	Starting-bars' (damper wdg.) condition (SP)			
R19	Bull-ring segments (loose, overheating?)			
R20	Starting-bars to bull-ring segment bracing (SP)			
R21	Collector rings' condition			
R22	Collector insulation condition			
R23	Brush-springs' pressure and condition			
R24	Brush rig condition (clean, damaged?, etc.)			
R25	Shaft-voltage discharge-brush condition			
R26	Inner/outer hydrogen seal condition (RR)			
R27	Circumferential pole slots' condition (RR)			

RR: Item to be found only in round-rotor machines
SP: Item to be found only in salient-pole machines

Form 6: Salient Poles Condition Report

Pole #	Condition	Pole #	Condition	Pole #	Condition
1		13		25	
2		14		26	
3		15		27	
4		16		28	
5		17		29	
6		18		30	
7		19		31	
8		20		32	
9		21		33	
10		22		34	
11		23		35	
12		24		36	

Form 7: Excitation Inspection

Type of excitation: *Self-excited* ____ ; *Stand-alone solid-state* ____ ; *Shaft-driven generator* ____ ;

Stand-alone generator ____ ; *Other* ____________________

"O" for satisfactory
"X" for unsatisfactory

Item	Description	N/A	O/X	Corrective Action
E01	Cleanliness (oil, carbon dust, etc.)			
E02	Shaft-mounted diodes condition			
E03	Diodes connections and support hardware			
E04	Commutator condition			
E05	Commutator brushes condition			
E06	Commutator brush-holder and rig condition			
E07	Brush-springs pressure & condition			
E08	DC generator stator condition			
E09	DC generator armature condition			
E10	Exciter-drive motor cleanliness			
E11	Exciter-drive motor stator condition			
E12	Exciter-motor rotor (squirrel cage, etc.)			
E13	Field discharge resistor condition			

Form 8: Comments

Comments: ____

Form 9: Wedge Survey

This is a typical table for performing a wedge survey. A larger number of columns for wedges and/or rows for slots may be required for larger machines. One way to enter the information is:

- "O" for a *tight* wedge
- "H" for a *hollow* wedge
- "L" for a *loose* wedge

SLOT #	WEDGE 01	WEDGE 02	WEDGE 03	WEDGE 04	WEDGE 05	WEDGE 06	WEDGE 07	WEDGE 08	WEDGE 09	WEDGE 10	WEDGE 11	WEDGE 12	WEDGE 13	WEDGE 14	WEDGE 15	WEDGE 16	WEDGE 17	WEDGE 18	WEDGE 19	WEDGE 20	WEDGE 21	WEDGE 22	WEDGE 23	WEDGE 24	WEDGE 25	WEDGE 26	WEDGE 27	WEDGE 28	WEDGE 29
1																													
2																													
3																													
4																													
5																													
6																													
7																													
8																													
9																													
10																													
11																													
12																													
13																													
14																													
15																													
16																													
17																													
18																													
19																													
20																													
21																													
22																													
23																													
24																													
25																													
26																													
27																													
28																													
29																													
30																													
31																													
32																													

Wedge Survey (continued)

SLOT #	WEDGE 01	WEDGE 02	WEDGE 03	WEDGE 04	WEDGE 05	WEDGE 06	WEDGE 07	WEDGE 08	WEDGE 09	WEDGE 10	WEDGE 11	WEDGE 12	WEDGE 13	WEDGE 14	WEDGE 15	WEDGE 16	WEDGE 17	WEDGE 18	WEDGE 19	WEDGE 20	WEDGE 21	WEDGE 22	WEDGE 23	WEDGE 24	WEDGE 25	WEDGE 26	WEDGE 27	WEDGE 28	WEDGE 29
33																													
34																													
35																													
36																													
37																													
38																													
39																													
40																													
41																													
42																													
43																													
44																													
45																													
46																													
47																													
48																													
49																													
50																													
51																													
52																													
53																													
54																													
55																													
56																													
57																													
58																													
59																													
60																													
61																													
62																													
63																													
64																													
65																													
66																													
67																													
68																													
69																													
70																													
71																													
72																													
73																													
74																													
75																													

Form 10: Electric Test Data

The following data is a sample of data from tests performed after machine shutdown (some of these tests to be performed under hydrogen for hydrogen-cooled machines)

refer to the pertinent standards for the correct test procedures — see references for chapter 8 —

MACHINE STATOR

- Measured conductivity for liquid-cooled machines: ______micro-mhos/cm
- Measured megger readings of windings to ground (with 2500 V megger) _______

  30 s _______ 1 min _______ 10 min _______
  Polarization Index (10 min/1 min) _______
  Ambient temperature _________ Hours after shutdown ________
  Stator RTDs temperature 1) _______ 2) _______ 3) _______ 4) _______
  5) _______ 6) _______ (two of each phase)
  Note: For water-cooled stators only 1 min megger reading required

MACHINE ROTOR

- Measured megger readings of windings to ground (with 500 V megger) _______

  30 s _______ 1 min _______ 10 min _______
  Polarization Index (10 min/1 min) _______

ALTERNATOR EXCITER

Stator

- Measured stator megger readings of windings to ground (with 500 V megger) _______

  30 s _______ 1 min _______ 10 min _______
  Polarization Index (10 min/1 min) _______
  Winding temperature _________

Rotor

- Measured megger readings of windings to ground (with 500 V megger) _______

  30 s _______ 1 min _______ 10 min _______
  Polarization Index (10 min/1 min) _______
  Winding temperature _________

DC EXCITER

- Measured megger readings of windings to ground (with 2500 V megger) _______

  30 s _______ 1 min _______ 10 min _______
  Polarization Index (10 min/1 min) _______
  Winding temperature _________

Electric Test Data (continued)

RTDs

- Megger test to ground with 500 V Megger
- Measure each RTD's resistance with a bridge
- Compare reading with measured temperature of the winding

STATOR WATER OUTLET THERMOCOUPLES

- Measure millivolts and compare readings with measured temperature of water or ambient air (when empty)

ADDITIONAL TESTS

- Corona probe (when required—large salient-poles)
- PD Activity readings (before shutdown if instrumented)
- Rotor-flux waveforms (if flux probe installed)

ALARM CHECKS

The following is a sample of the alarm circuits and activators that require check (different machines will have a different set of alarms):

- Air filters clogging alarms
- Stator cooling water pressure low
- Water pressure at machine
- Water flow
- Stator water filter
- Stator water-cooling pump
- Water and oil leakage detectors
- Hydrogen seal-oil enlargement detector
- ______________________________
- ______________________________
- ______________________________

PART 2
Inspection

Chapter 4 **Description of Stator Items**
Chapter 4 describes each stator item on Form 4, Stator Inspection.

Chapter 5 **Description of Rotor Items**
Chapter 5 describes each rotor item on Form 5, Rotor Inspection.

Chapter 6 **Description of Excitation Items**
Chapter 6 describes each excitation item on Form 7, Excitation Inspection.

Chapter 7 **Generator Auxiliaries**
Chapter 7 describes auxiliary systems that should not be overlooked in an inspection, such as lubrication systems, hydrogen seal oil systems, stator-cooling water systems, and hydrogen systems.

Chapter 8 **Standard Electrical and Mechanical Tests**
Chapter 8 gives an overview of the electrical and mechanical tests, which should only be performed by trained personnel.

4

Description of Stator Items

The Stator Inspection form refers to items comprising the actual stator, as well as the frame, bearings, and other machine-related components. Each item on the form is described below with reference to its item number. Figure captions also include the reference number for the item to which they correspond.

S01: Cleanliness of Bore

Important information on the condition of the machine may be obtained from a general view of the bore area and frame. For instance:

- Excessive discoloration (and perhaps flaking) of paint on the casing, frame, and bore indicates a probable case of overheating. This could be a result of overloading, or improper flow of air, gas, or water.
- The presence of large amounts of oil or a dust-oil mixture attests to possible hydrogen-seal problems.
- In certain types of air-cooled machines, large amounts of carbon dust are evidence of deficient sealing between the collectors' enclosure and main bore areas.
- In cylindrical machines (turbine generators), loose copper dust, or dust mixed with oil and/or other dusts, indicates excessive pounding of the rotor-field conductors.

- Water found in the bottom of the machine may indicate a leak in the heat exchanger.
- Excessive amounts of iron powder mixed with oil and dust or found alone in the bore area tend to indicate a loose core.

Nuts, bolts, small pieces of lamination iron, or other loose objects found inside the machine, oftentimes at the bottom of the casing, should be investigated as to their origin. They may point to loose space heaters, broken laminations or cooler fins, and/or other abnormalities in need of attention.

Figure 4-1 shows metallic objects found in the bore and bottom of the casing of an air-cooled, gas-turbine generator. In order to retrieve these objects, side plates were opened and long instruments with a grip device at their end were used. Subsequent examination of the bore area identified these metallic objects as remnants of broken pieces of laminations (see Fig. 4-7).

Fig. 4-1 [S01] Broken pieces of laminations found in the bore area and bottom of the casing of an air-cooled gas-turbine generator.

S02: Air/Gas-Ducts Clogged/Unclogged

Clogged vents effectively derate a machine by restricting the flow of cooling gas through the laminations and coils. This phenomenon is particularly evident in open-air machines, which tend to be older and slower: mainly hydrogenerators, condensers, and industrial motors. Clogged vents are particularly common in open-air machines contaminated with oil.

The restrictions can show up during operation of the machine as hot spots; i.e., temperature readings from one or several temperature detectors will be higher than those obtained from the rest. (This can also indicate core-insulation problems, as discussed below.)

The clogging need not only be dust or show itself only in old or open machines. In one case, massive clogging of the ducts by red iron oxide powder was found in a new large hydrogenerator having a very loose core.

Normally, a visual inspection with the aid of a frontal source of light may suffice. Where possible, a light placed in the back of the core while observing from the inside can result in good observations of the ducts. In cases where the rotor has been left in place and clogging of the ducts is suspected, insertion of a side-view boroscope through the airgap can provide an excellent view of the air ducts in the inspection area.

S03: Iron Oxide Deposits

Iron oxide appears as red powder deposited mainly on the bore and in the air/gas-ducts of the machine. When mixed with oil, these deposits may be concealed in a mixture of dirt and oil. This mixture should be chemically analyzed when iron dust is present (or suspected), for content in proportion to weight. A quick identification during the inspection can be made by subjecting small portions of the mixture to the field of a magnet (one of the desirable inspection tools). If the "dirt" responds to the magnetic attraction, then iron dust is most certainly present in significant proportions.

Iron oxide can result both from loose laminations and loose wedges [1]. In the case of loose wedges, the iron dust is mainly seen in the contact region between wedge and iron. In the case of loose laminations, the iron oxide powder deposits are distributed on larger sections of the machine, on the iron itself. The deposits will tend to concentrate in places adjacent to the air/gas-ducts, having been left there by the flowing gas (see Figs. 4-2 and 4-3). In severe cases, the large amounts of iron oxide powder generated might clog air/gas-ducts.

When the amounts of iron oxide powder present in the machine are substantial, its origins should be thoroughly investigated. If loose wedges are the cause, then they should be tightened by re-wedging or another effective method. Loose wedges may abrade themselves to the extent they come out of the slot (see Fig. 4-4). They may also indicate a loose winding condition, with detrimental consequences to coil insulation; in particular, the loss of semiconducting paint [2, 3].

If the iron oxide originates from the movement of metallic parts, it probably indicates a loose core or loose portions of the core [4].

Cores are pressure-loaded to given values during the manufacture of the machine. During operation, the core is subjected to continuous elongation and contraction of the laminations (magnetostriction) at twice supply frequency, and

Fig. 4-2 [S03] Section of a large turbine-generator bore showing deposits of red iron oxide powder.

Fig. 4-3 [S03] Close view of Figure 4-2. The accumulation of the red powder depends on the origin of the powder and the pattern of the flow of cooling gas.

Fig. 4-4 [S03] Wedges in the stator of a synchronous condenser worn due to vibrations to the extent that several came out of the slot, and others were found to be on their way out.

elongation and contraction of supporting structures due to thermal cycles and vibrations. Machines having properly designed and stacked cores are supposed to withstand these onerous conditions. In many cases, however, after years of operation, this constant movement of the core components and the accompanying metal fatigue, abrasion, and deformation result in a reduction of the core's loaded pressure. The end result, if not corrected, is abrasion of the interlaminar insulation (Figs. 4-5 and 4-6). The consequences of such abrasion are spot-heating of the core, broken laminations leading to machine contamination with iron particles, and serious failures of core-compression bolts, bolt insulation, and core-compression fingers [4]. Other detrimental effects are deterioration of the coil insulation due to hot spots in the core, additional core losses, and augmented vibrations and increased audible sound levels. An additional test to confirm the presence of a loose core is the insertion of a knife between the laminations at several locations. If a 10-mil blade penetrates more than a quarter of an inch, the core may not be sufficiently tight. Extreme care should be taken not to break the blade, leaving a piece in the laminations. This technique should be used very carefully, especially in any machine with its windings in place. (For a description of the proper procedure, see Reference [3].)

Some utilities check the torque of a sample of the machine's compression bolts at every second or third overhaul, or some other chosen interval. The measured torque value is compared with those recommended by the machine's manu-

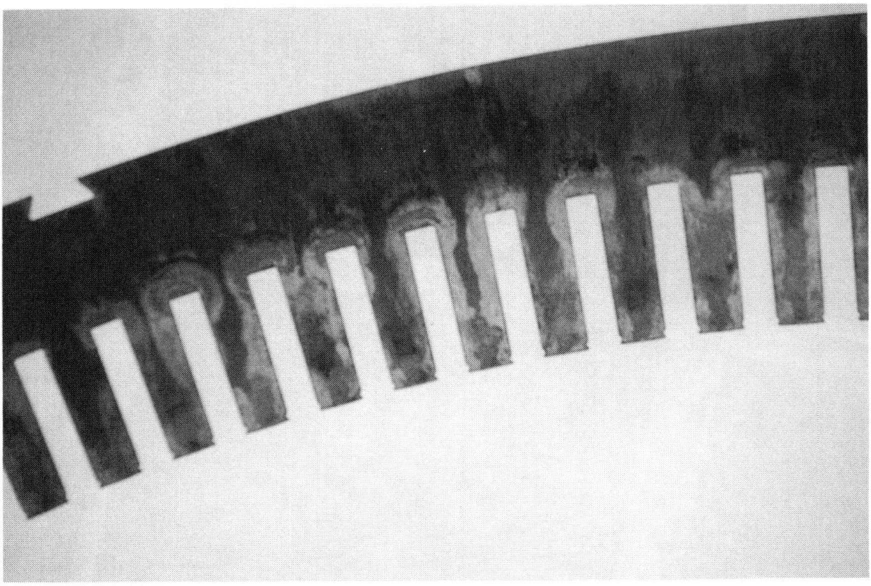

Fig. 4-5 [S03] Section of stator laminations showing the insulation partially lost by abrasion.

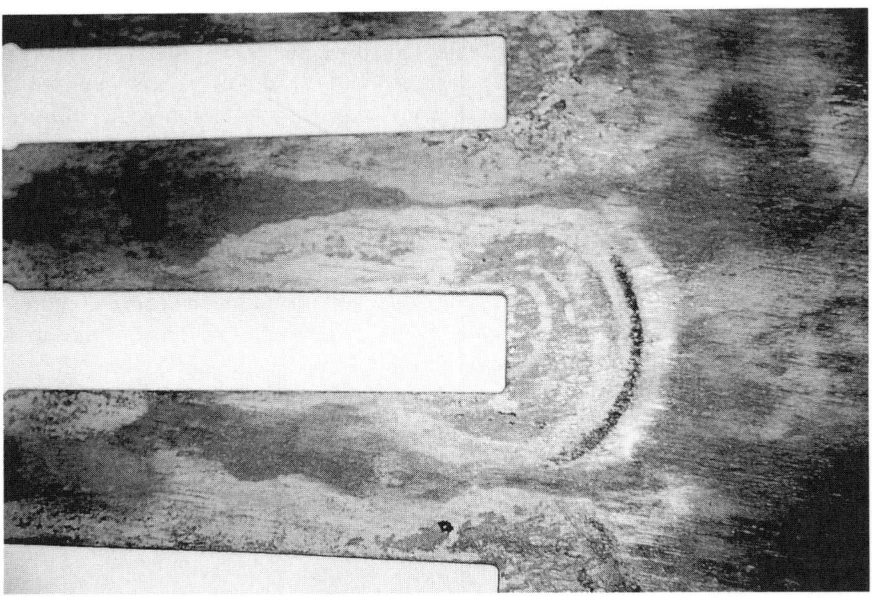

Fig. 4-6 [S03] Close-up view of a section of the lamination shown in Figure 4-5, with the lamination's insulation abraded.

facturer. If these bolts are found to be loose, the entire core is retorqued in accordance to procedures laid down by the manufacturer.

It is important to note that not all lamination problems are the result of a loose core. For instance, Figure 4-7 shows entire packets of broken laminations because of lack of sufficient support by the I-shaped duct separators at the top of the tooth. The core of this particular machine (an air-cooled gas-turbine generator) was otherwise found to be tight. In this case, repairs included the introduction of epoxy glass laminates between laminations, and application of penetrating (low-viscosity) epoxy to the damaged area.

The solution to loose laminations is varied, depending on location, severity, and type of problem. It ranges from retightening the core-compression bolts to changing compression plates or parts thereof, introducing nonmetallic shims, and so forth.

Fig. 4-7 [S03] Two packets of broken laminations belonging to an air-cooled gas-turbine generator.

S04: Hardware Condition

All parts in a generator are exposed to continual vibration, temperature changes, and other mechanical stresses. They may become loose, fractured, or broken. It is therefore important to search for these abnormalities during the inspection before they develop into major troubles.

In particular, all components of the winding support assembly are subjected to mechanical stresses due to sudden and large load changes, such as are present during loss of load, short circuit, pole slipping, and closing out of synchronism. Machines subjected to the above-mentioned conditions, as well as machines operated with many starts, are more susceptible than others to hardware failure.

Some of the most sensitive components are:

- Compression bolts—Observe if any *greasing* (oil and dust mixed together by the friction of two components vibrating within the machine) is present indicating loose bolts or core, and integrity of nut-locking device.
- Surge-ring supports—Look for cracking and looseness.
- Finger-plates—Look for cracked or bent fingers.

Observe space heaters for looseness of bolts and nuts, and integrity of connections. Figures 4-8 and 4-9 present an example of hardware in the bore of the machine damaged during removal of the rotor.

Fig. 4-8 [S04] Section of a large 4-pole turbine generator showing gas-guides made of insulating material.

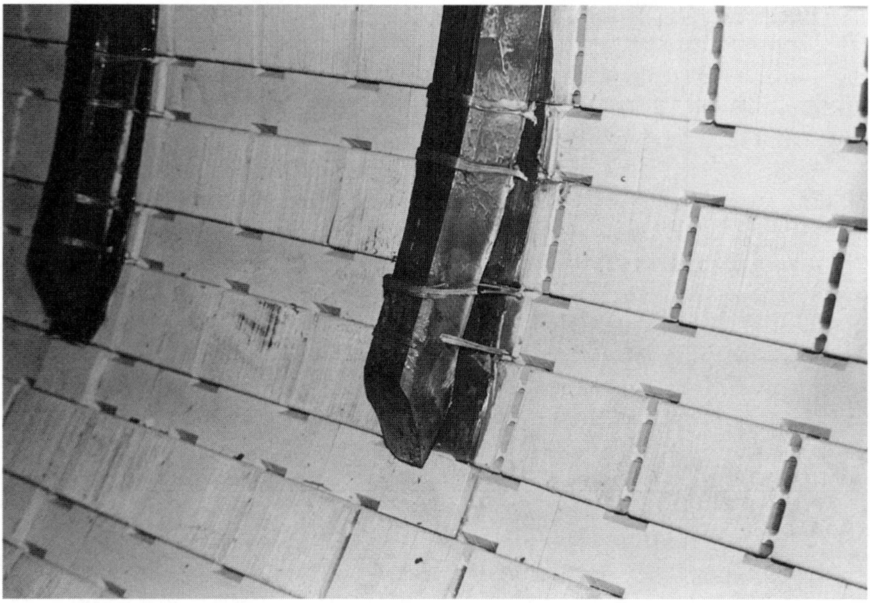

Fig. 4-9 [S04] Close-up view of the gas deflectors showing damage during the removal of the rotor (see Fig. 4-8).

S05: High-Voltage Bushings

There are too many different arrangements of terminal boxes and bushing types in large synchronous machines to describe them all in this book. However, the following general inspection guidelines and comments apply to any arrangement.

Lead-bushings are susceptible to damage arising from sudden load changes, excessive vibration, overheating of the leads, and normal vibration over long periods of time. Stator high-voltage bushings should be inspected for evidence of cracks, oil leakage (when oil-filled), and looseness of components. All dirt and tracking residues should be thoroughly cleaned.

In large turbogenerators, the high-voltage bushings are partly contained in sealed bushing wells. Some of the lead-bushings have ducts allowing the flow of hydrogen. The ducts should be free of oil, grease, or any foreign elements. Many others are water-cooled. In these, connections to the bushings should be inspected for cracks and leaks.

S06: Stand-Off Insulators

As with high-voltage bushings, stand-off insulators are subjected to continuous forces due to vibration, as well as thermal expansion and contraction of the leads (see Fig. 4-10).

Stand-off insulators should be inspected for cracks and looseness of their constituent components. External surfaces should be kept clean to avoid extensive tracking, which may result in eventual short circuits to ground.

Old insulators may utilize lead in their construction. In many cases, continuous vibration over a long period of time will deteriorate the insulator, to the extent that the lead-based material will ooze out of the insulator in the region of the porcelain seal. This appears as a gray powder or paste.

Stand-off insulators are found inside the case and bushing well of large turbogenerators, as well as supporting the lead and neutral busses of all other types of machines. In some instances, the busses may extend to significant distances (sometimes below) from the main machine body. They should be inspected for cleanliness and overall condition.

It has been widely documented that there is a relationship between failures in the connecting leads of large turbogenerators and the condition of the stand-off insulators. In many cases, in large 2- and 4-pole machines in which failures of the connection leads occurred, the stand-off insulators showed moderate to heavy signs of "greasing." The greasing is the result of the dust produced by excessive

Fig. 4-10 [S06] Stand-off insulators damaged because of surface contamination.

fretting of the seal between the porcelain and the mounting flange, and its mixture with oil or oil vapors present in the bushing well.

The normal mode of lead failure is fatigue cracking of the flexible connections due to poor support of a weakened stand-off insulator. Partial cracking of the flexible connections results in increased temperature of the conductor. Eventually the connectors melt, resulting in serious failures.

Additional serious damage can result from the change in the cooling path of the hydrogen due to a shift in the position of the porcelain of the insulator in relation to its supporting flange.

S07: Bushing Vents

As explained in item S05, bushings sometimes have passages built inside them to allow the flow of air or hydrogen for cooling purposes. It is important that the vents are inspected to see if they are unclogged. In sealed bushing wells, high-voltage bushings may be flooded by seal oil, and their passages clogged or semi-clogged. If clogged, they can be siphoned clean with a venturi pipe or vacuum. Figure 4-11 shows a cross section of a typical bushing well.

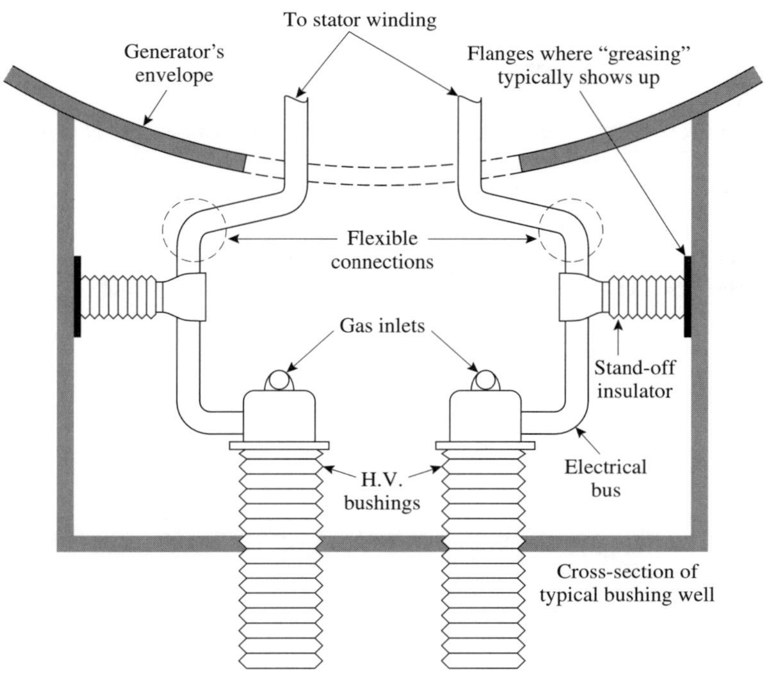

Fig. 4-11 [S07] Cross section of a typical bushing well showing stand-off insulators and their gas passages.

When possible, a nipple (or nipples) can be placed at the bottom of the bushing well, protruding about 1/2 inch inside the bushing well. The nipple's protrusion allows the retention within the well of the required viscous-oil seal normally placed in these bushing wells of hydrogen-cooled machines. However, it will allow the flow into a detection and purging pipe of unwanted water and seal oil from the shaft seals that may leak into the machine and overflow the protruding length of the nipple.

S08: Greasing/Red Oxide Deposits on Core Bolts

In large machines, the stator core is almost always compressed by means of *core-compression bolts*. These are torqued to values specified by the manufacturer during the assembly of the machine (normally to around 150 psi of core pressure). Over time, part of the initial pressure within the core is lost. This looseness of the core is indicated by movement of the nuts and washers on the core-compressing bolts against the compression plates. This movement results in red iron oxide deposits or, when oil is present, in a greaselike substance.

If these greasing and/or red iron oxide deposits are encountered during an inspection, the core-compression bolts should be tightened to the manufacturer's specified values.

In the case of hydrogenerators, it is common for the manufacturer to call explicitly for retightening of the core-compression bolts after one year of operation of new machines.

Failure to tighten the core to specified values may result in core damage, as well as damage to the insulation of the core-compression bolts (refer to item S03, above).

Following is a simple method of calculating the torque required on the core-compression bolts of many-pole machines. Typically, these are hydrogenerators, rotary condensers, and slow synchronous motors.

The first step is to calculate the net radial area of the core (see Fig. 4-12).

$$\text{Net Area}\left[\text{in.}^2\right]$$

$$= 0.7854\left(\text{OD}^2 - \text{ID}^2\right) - \left(\text{Number of slots} \bullet \text{width of slot} \bullet \text{depth of slot}\right)$$

(all dimensions in inches)

$$\text{Total force in core}\left[\text{lb}\right] = \text{Net Area}\left[\text{in.}^2\right] \bullet 150\,\text{lb}/\text{in.}^2$$

The 150 lb/in.^{2} is a typical value. Core pressure values for specific machines can be obtained from the manufacturer.

$$\text{Force per bolt "F"}\left[\text{lb}\right] = \left(\text{total force in core} \bullet \text{D}/\text{d}\right) / \left(\text{number of bolts}\right)$$

The required torque per bolt is:

$$T\left[\text{ft-lb}\right] = \text{Force per bolt}/\text{bolt torque constant}$$

The bolt torque constant is a number that depends on the type of the bolt and bolt size. It is tabulated by bolt manufacturers and the ASTM (American Society for Testing Materials) Std A-193-B7 Chart.

The following formula can also be used to calculate the torque values from the axial load on the bolts, if enough information is available on the bolts:

$$T\left[\text{ft-lb}\right] = \frac{F \bullet Dm}{2} \bullet \left[\frac{\cos\phi \bullet \tan\lambda + f}{\cos\phi - f \bullet \tan\lambda}\right] + \frac{F \bullet f \bullet Dnut}{2}$$

where:

F = total axial force in bolt [lb]

Dm = mean pitch diameter of bolt [in.]

$Dnut$ = mean nut diameter [in.]

= (minor diameter + distance across flats)/2

ϕ = angle of thread

f = coefficient of friction (from manufacturer's tables, standards, etc.—depends on angle of thread and dry or wet thread)

$$\tan\lambda = \frac{\text{Lead}}{\pi \bullet Dm}$$

Lead = distance between threads [in.] (for double threads, one thread is skipped in the measurement of the distance between the threads)
Refer to ASTM A-193-B7 for the required values.

In 2- and 4-pole machines, the compression bolts travel through the core laminations. In this case, the force per bolt is approximately equal to the Total force in core, divided by the number of bolts. For accurate torque values, always contact the respective manufacturer.

It is very important to recognize that in machines with expansion-bearing-bolts which allow the end-windings to slide (see item S41), the end-windings may have to be relieved from their brackets before the core is compressed. Subsequently, after the core is compressed to the desired pressure, the end-windings are tightened again to the supporting brackets. This operation normally entails releasing and tightening several bolts per bracket, which allows the two halves of the brackets to separate. The reason for loosening the bracket halves before compressing the core is to avoid building tension on the end-windings when the core shrinks during the core-compression operation. The same may apply to machines with long cores and without expansion-bearing-bolts. Consult with the manufacturer when in doubt.

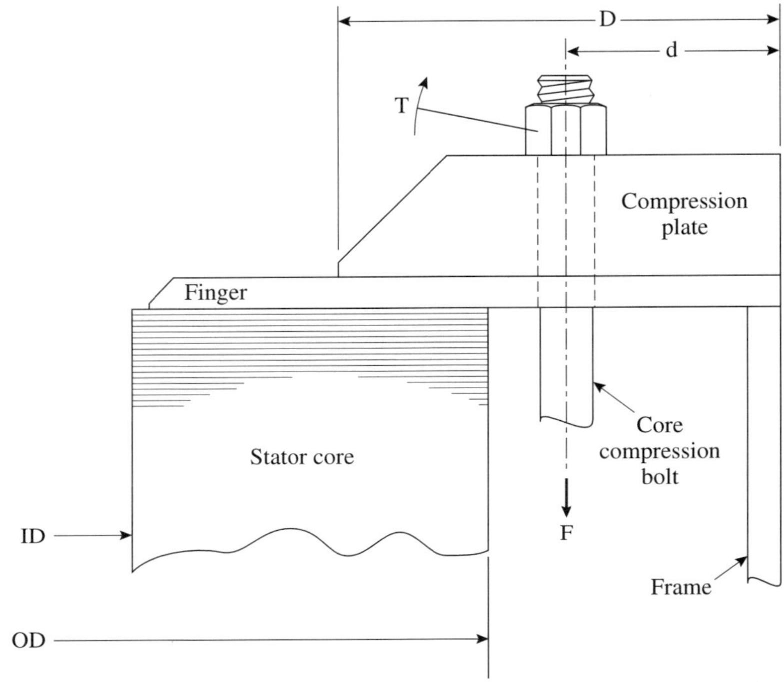

Fig. 4-12 [S08] Schematic arrangement of the core-compression hardware of a typical many-pole machine.

Bell-Shaped (Spring) Washers. In certain machine designs, the connection rings are kept tight by means of "bell"-type spring washers. They are conic-shaped, and they are inserted at both ends of the core-compression bolts (see Fig. 4-13).

Bell washers have been known to break by a process known as metal fatigue. Being located at the end of the core, in the region of high-velocity moving gas/air created by the rotor fans, the broken parts can be thrown with force onto the end-windings. This can result in immediate failure or partial damage to the insulation with potential for later development into catastrophic faults of the insulation.

One common practice that eliminates this concern is the use of resin-soaked glass tape to wrap the washers together with the bolt heads and/or nuts. In this manner, all broken parts of the spring washer will remain contained within the hardened glass tape.

During major overhaul inspections (which should be done every 5 to 8 years), it is good practice to remove the glass tape from the core-compression bolt ends and bell washers to allow their inspection and assess the condition of

Fig. 4-13 [S08] Spring washers covered with resin-soaked glass tape.

the washers. There are some users who like to change these washers at intervals of several years (during a major overhaul) even when they don't show visible signs of deterioration.

S09: Space Heaters

It is common with many large synchronous motors and generators to have a number of space heaters installed. Very often they are located on both sides of the frame, at the end-winding region.

The function of the space heaters is to keep the moisture out of the machine during periods of no operation. By warming the internal air or gas, the water vapor always present in the air or gas does not condense on the windings. Water condensation is a major enemy of insulation. Thus, keeping the space heaters in good operating condition is an important element of correct maintenance when operating large electric rotating machinery.

The inspection should, as a minimum, check for loose connections and integrity of the wires and heating elements. If visual access is not satisfactory, then a resistance test taken at the terminal box should be performed to ascertain continuity of the circuit.

S10: Fan-Baffle Support Studs

Most large machines include air or gas baffles to direct the cooling air or gas to and from the fans. In this manner a cooling circuit is established in the machine.

The so-called *fan-baffles* are subjected to continuous and diversified modes of vibration. Studs or bolts should be inspected for stress-fatigue cracks (see Fig. 4-14). If broken during operation, they will very probably cause extensive damage to the windings or rotating elements, particularly the fan blades.

Fig. 4-14 [S10] Fan-baffle support threaded stud shown. The stud is mounted on the tip of the core-compression finger. A number of equally spaced studs are placed around the bore.

S11 and S12: Heat Exchanger Cleanliness and Leaks

Heat exchangers are an integral component of the machine. Typically, large turbogenerators have their hydrogen–water or air–water heat exchangers located inside the machine's casing. In slower air-cooled machines, the air-to-water heat exchangers are located in a separate containment, often below the machine. In all cases, the maximum allowable rating is directly dependent on satisfactory condition of the heat exchangers.

The two major concerns with heat exchangers are:

- clogged water tubes
- water leaks

Through continuous flow, water tends to clog the ducts by corrosion and deposits of minerals. Once initiated, the clogging process advances at an increased pace. Not only is the flow of cooling water reduced, but also the heat transfer coefficients are detrimentally affected due to the mineral buildup inside the pipes, resulting in poor heat-transfer properties.

It is therefore important to visually inspect the tubes and recommend corrective action (such as descaling) when required.

Heat exchangers, in particular those positioned inside the machine's case, are continuously subjected to vibration. Therefore, it is not uncommon to find leaks developing during the operation of the machine. There are several methods available for the detection of those leaks. Hydrogen sensors are commonly used in hydrogen-cooled machines. The higher pressure of the hydrogen creates a flow of the gas into the coolers, keeping the water from entering the machine. Even in hydrogen-cooled machines, as in other types of machines, water heat-exchanger leaks will probably result in water contamination of the windings. A visual inspection can detect a leak by looking at suspicious accumulations of water inside the casing. Pressure tests are also routinely performed, and are recommended for coolers installed inside the machine that haven't been inspected for several years. A leak can also be checked by opening the drain tap normally located at the belly of the machine and looking for water. Tubes can also leak from thinning of their walls by erosion and corrosion. Eddy-current tests can detect excessive thinning of the walls.

S13: Hydrogen Desiccant/Dryer

A typical maximum dew point for hydrogen in a machine is 30°F (–1.1°C). Periodic purity checks should suffice to maintain the dew point low enough to avoid condensation during operation of the machine. Nevertheless, if purity cannot be assured, it is then important to ensure that only dry hydrogen enters the machine. Moisture carried by the hydrogen into the machine, if allowed to condense, may adversely affect critical components.

Moisture in the windings can cause tracking with subsequent failure of the insulation. Moisture on the retaining rings can be a serious source of problems, in particular if the rings are made of austenitic steel (18% manganese-5% chromium alloy), commonly known as *18-5 rings.* In the presence of moisture, this material tends to develop stress-corrosion cracking, which grows relatively fast, and this can result in catastrophic failure of the ring. Retaining rings made out of 18-5 alloys are being changed throughout the industry for 18-18 alloy rings, following recommendations by the manufacturers. Some machine operators keep their 18-5 rings but subject them to periodic NDEs and are very careful to avoid ingression of moisture to the machine. The new "18-18" rings have proved to be less prone to stress-corrosion cracking, but are still susceptible to water-initiated corrosion. Moisture carried by hydrogen can also result in corrosion of the iron laminations, zone rings, fan rings, nonmagnetic rotor wedges, bolts, studs, brakes, and other structural members. Carbonate byproducts from welding fluxes used in hydrogen coolers have been known to result in winding failures [5].

However, manufacturers are divided as to the best methods to maintain dry hydrogen. Most manufacturers base their decision for installing or not installing hydrogen dryers on their bearings' oil-seal arrangement. Some manufacturers do not recommend the use of hydrogen dryers [5].

The approach suggested here is that if hydrogen dryers are installed, they should be inspected to ascertain they are in good working condition. Desiccants are often neglected, particularly in places where scheduled maintenance procedures are lacking; therefore, they should be visually inspected as part of the machine inspection procedure. However, plant personnel should make sure hydrogen dryers are in good working condition at the time of shutdown. This is the most critical time, because it is when the moisture will condense on critical components of the machine.

S14: Core-Compression (Belly) Belts

Many types of 2-pole machines include *compression belts,* also known as *belly belts,* around their spring-supported stator core [6]. These provide for control of the vibration levels of the machine. It is not uncommon for additional compression belts to be installed if core vibrations reach unsatisfactory levels.

Belly belts are normally not accessible during an inspection. Their inspection is rarely considered other than when stator vibration problems are present and can be attributed to insufficient peripheral compression of the core-suspension arrangement. Manufacturers can provide the values to which compression belts should be torqued.

S15: Bearing Insulation (at Pedestal or Babbitt)

During the normal operation of electrical rotating machinery, voltages—and hence currents—are induced in the shaft. In the case of large turbine generators, in addition to those voltages induced in the generator, substantial voltages are created by the rotating elements of the turbine. These shaft voltages, and in particular the resultant currents, have to be kept to low values; otherwise, subsequent bearing failures can be expected to occur.

The main sources of shaft voltages are [7]:

- *Potential Applied to the Shaft*—Applied intentionally or accidentally. For instance, a grounded field-conductor and excitation spikes.
- *Electrostatic Effects*—Due to impinging particles (e.g., shaft-mounted fans in generator; turbine blades in steam or gas turbines) or charged lubricants.
- *Dissymmetry Effects*—Due to change of the reluctance path as a function of the angular position of the rotor due to magnetic asymmetries of the core. (This can be due to design, manufacturing details, or to core faults of large magnitude.)
- *Homopolar Flux Effects*—Due to axial fluxes originating from magnetized turbine and generator components, especially the generator's shaft.
- *Movement Off Magnetic Center*—Due to axial movement of the rotor off the magnetic center, which might result in shaft voltages. It has been noticed that many-pole hydrogenerators are more sensitive to this phenomenon than 2- and 4-pole generators.

In large turbine generators, voltages of up to 150 V peak-to-peak are not uncommon [8]. These voltages have the potential to generate currents that, when allowed to flow freely, will destroy bearing surfaces, oil seals, and other close-tolerance machined surfaces.

The damage is not caused by the heat developed by the flow of current, as sometimes assumed; rather, it is mechanical. Shaft bearing currents will damage the bearing surfaces by pitting resulting from minuscule electric discharges. The pitting will continue until the bearing surfaces lose their low coefficient of friction; then other more dramatic and fast changes occur, culminating in bearing failure. Bearing currents also have adverse effects on the lubrication oil by altering its chemical properties.

Control of shaft voltages is achieved by taking certain precautions when designing the machine and by introducing shaft-grounding elements and/or bearing insulation.

Grounding devices, which can be copper braids, silver graphite, or copper graphite brushes, should be inspected often. Bearing insulation can be inspected often with certain bearing-insulation designs (with the machine on-line), or with the machine off-line, and after a certain amount of disassembly of pipes and other attachments has been performed.

During major inspections, the bearings are inspected for signs of pitting caused by shaft (bearing) currents. These are easily recognized with help of a magnifying glass. They appear as shiny and well-rounded little droplets. If the bearing surfaces are found to be damaged by shaft currents, then the reason for the existence of significant shaft currents should be investigated. Strong candidates are a bridged or faulty bearing insulation, and/or defective or missing grounding devices. Pitted surfaces of the bearings should be rebabbitted. Pitted journals should be polished.

The bearing insulation is commonly made of mechanically strong water- and oil-resistant insulation materials formed from fiberglass, or similar bases, laminated and impregnated with resins, polyesters, or epoxies.

In bracket-mounted bearings, the bearing insulation takes the shape of a ring or collar surrounding the bearing or, less commonly, as an insulation layer between bracket and casing, piping, and so forth. In pedestal-type bearings, the insulation is made of one or two plates, normally placed at the bottom of the pedestal. When double-plates are used, they "sandwich" between them a grounded metal plate. This system allows each of the insulation layers to be tested without the necessity of interrupting the operation of the machine, uncoupling it, and taking care of the other arrangements normally required to measure the bearing insulation.

Machines normally have the non-drive-end (outboard end) bearing insulated. However, machines with couplings at both ends or driven by turbines may have both bearings insulated. In some cases, the couplings are also insulated.

It is important to verify during the inspection that the bearing insulation is not contaminated with carbon dust or accidentally bridged with chunks of metal touching the bearing pedestal, temperature or vibration sensors, noninsulated oil piping, and so forth.

If electrical testing of the bearings' insulation is performed, the measured values should be in the hundreds of thousands of ohms. However, in this application, as in many other areas of insulation practice, a wide range of resistance values can be found in the literature: from 100 kΩ to 10 MΩ or greater. Given the relatively large shaft voltages encountered in large turbine generators, it is preferable to have insulation-resistance values in the MΩ region.

Reliance only on grounding devices is not recommended. A grounding device is required to reduce the level of voltages in the shaft to values compatible with the small clearances encountered in bearing insulation and between shaft and seals. However, the normal contact resistance of grounding devices does not eliminate shaft voltages to the extent that bearing insulation would become redundant [9].

Grounding devices are often taken for granted—and therefore neglected—during normal maintenance procedures. Thus, it is important they are inspected carefully during major inspections.

S16: Coil Cleanliness

It is almost certain that during most comprehensive inspections some contamination will be found on the coils of the machine. Given that these intensive inspections are performed at intervals of several years, and given the adverse effects contamination has on the integrity of the insulation, it is appropriate to recommend cleaning the coils.

Contamination causes degradation of the insulation by providing a medium for currents to flow on the surface of the coil insulation. This results in tracking and reduction of the insulation properties. In addition, contamination may penetrate the insulation cell via cracks, which occasionally results in tracking followed by a short circuit.

Normally, the inspector will determine the measures required for cleaning the insulation based on the degree of contamination found during the inspection. There are several choices; some are mechanical, such as vacuum, compressed air, crushed corn cobs, lime dust, CO_2 pellets, brushes, and cloths. Chemical methods include solvents. Steam cleaning is another option.

The choice of any particular cleaning method should be based primarily on the type of winding insulation, the degree of dirt or contamination encountered, the condition of the windings (i.e., how they might be affected by the impact of particles, steam, etc.), and how exposed they are to solvent ingression [10]. Various states and jurisdictions may impose different regulations on the use of the many solvents available. Reference [4] (Item 7, page 11) contains a well-written and detailed description of different methods for cleaning windings.

Water-Cooled Windings. Water-cooled windings present a particularly severe problem regarding oil and/or water contamination. The design of the coil support system within the slots, and the higher electromagnetic radial forces present in these machines, make these types of windings less tolerant of oil and/or water contamination.

Oil contamination of the ripple (spring) fillers located on one side of the slot reduces the friction designed to retain the coil bars under the large radial forces. A winding too saturated in oil contamination could be beyond easy repair and require a full rewind.

Water ingression is damaging to the wall insulation of the bars. Control of water and oil ingression is primarily an on-line monitoring activity. However, if excessive oil and/or water contamination is encountered during a major inspection, the inspector has to evaluate the type of response required to return the machine to a reliable operating condition.

S17: Blocking Condition

The *blocking* consists of the material (items/inserts) used to separate the sides of the coils at the end-windings, and between end-winding connections.

Design trade-offs determine the clearances between the coil sides at the end regions, which are, in general, relatively small (of the order of a fraction of an inch). Tolerances of the manufacturing process of the coils, as well as mechanical stresses arising from the bending and pulling of the coils during their installation in the machine, result in many coils touching each other at the end-windings. Left alone, the coils will rub each other at twice supply frequency during the operation of the machine. Moreover, during starting, sudden changes of load (or external short circuits in the case of generators) create large movements of the coil ends. The continuous movement, combined with the rubbing between coils and the large sudden movements during starting and large load changes, as well as different temperature-related expansions and contractions, are strongly detrimental to the integrity of the wall insulation of the coils and can severely reduce the expected life of the windings. To eliminate the movement between coils, blocking is used.

Solid blocking material is almost always held in place with ties (see Figs. 4-15 and 4-16). In such cases, the condition of the ties determines how effectively the blocking performs.

To effectively eliminate the rubbing between neighboring coils, and to minimize end-winding movements during starting or large load changes, end-windings are restrained with surge-rings, ties, and blocking.

The blocking is carried out by inserting small separators between the coil sides. They come in two basic forms: solid or amorphous blocking. Solid blocking can be of rectangular shape or conform to the shape of the coils. Both shapes are made of solid insulating materials, such as textolite, maple, and so on. Amorphous blocking is made of felt or feltlike material, which is soaked in resin or equivalent

Fig. 4-15 [S17] The end-winding of a turbogenerator. Shown are the ties between the coils.

Fig. 4-16 [S17] Close-up of a blocking arrangement shows resin-soaked and hardened blocking made of felt and ties.

impregnating material during installation, or as part of a vacuum-pressure impregnation (VPI) process.

Feltlike material is found in most modern machines, while older windings are primarily blocked with solid separators. Often, solid separators tend to get loose after long periods of operation. In severe cases they will even fall from the winding. Operation of the machine with loose or missing blocking will adversely affect the reliability of the winding. Therefore, loose or missing blocking should be treated or replaced as necessary.

Inspection of the end-windings should include evaluation of the condition of the blocking. Signs of loose blocking include: greasing; dry, loose, or broken ties (see next item); powder; abrasion signs on coils; and missing blocking pieces.

Suggested corrections can include one or more of the following: retying; new blocks and ties; amending the coil wall insulation; cleaning and applying penetrating epoxies or resins.

S18 and S19: Ties between Coils

Blocking made of solid materials is secured with ties. These are made of glass roving, glass cords, flax, or other similar materials. They are impregnated with resins to form a solid tie holding together two coil sides at the end-windings, with the blocking snugly fitted in between. When exposed over a long period of time to the relatively high temperatures encountered in the machine, ties tend to "dry" by evaporation of the solvents and become brittle. Thus, their structure is deteriorated and they lose volume; becoming loose, they allow relative movement between the coil sides. Subsequently, the ties and blocking are further deteriorated by abrasion, with the negative consequences to the winding integrity detailed previously in item S17.

Ties are therefore an item to be inspected carefully. Defective ties may exhibit powder deposits, flaking or tearing signs, or other telltale signs of deterioration. If relatively few ties show an inadequate condition, they can be treated with lightly viscous epoxies to fill the voids, therefore tightening anew the tie-block structure upon drying out. If a substantial number of ties show signs of degradation, retying the complete winding should be considered.

S20 and S21: Ties to Surge-Rings

As explained in item S17, the end-windings of electric machines are subject to substantial movement during sudden changes of load, in particular during short circuits in the "electrical vicinity" of the machine. These movements are detrimental to the integrity of the insulation. Severe movement will deform the windings and possibly crack the insulation of coils, lead connections, coil connections, and so on. To minimize the damage inflicted by end-winding movements, the side coils are tied to circular rings commonly called *surge-rings* or *support-rings.* In large machines the rings are normally made of steel and sometimes of fiberglass materials. The steel rings are themselves covered with several layers of insulation.

Ties securing the end-windings to the surge-rings suffer the same type of temperature- and abrasion-dependent degradation as the ties between coil sides. When this occurs and it is allowed to persist for extended periods of time, it may result in abrasion of the wall insulation of the coil and the surge-ring insulation. The natural consequence is a phase-to-phase or ground fault at this location.

As with the blocking ties, the inspector should look for signs of excessive dryness, greasing, deposits of powder (normally of bright color), and so forth. The repairs are similar to those proposed for the blocking ties (items S18 and S19).

S22: Surge-Rings' Insulation Condition

The purpose of the insulation on surge-rings made of steel is to minimize the possibility of a ground fault to the rings. As described in the preceding item, the movement between coils and surge-ring is minimized by effective ties, but not completely eliminated. The surge-ring insulation provides an additional layer of protection in a critical area of the machine.

Inspection of the surge-rings' ties and insulation more often than not requires mirrors, since they are usually in a restricted place, not allowing direct visual access. However it is done, inspection of the windings should always include evaluation of the integrity of the surge-ring insulation, in particular in the areas beneath the ties. Greasing, powder deposits, and other telltale signs should focus the inspector's attention to a probable degraded condition. In addition to deterioration due to movement of the coils, the surge-ring insulation can deteriorate due to electrostatic discharges from the coils. These appear as electric tracking and/or burnlike marks on the insulation of the surge-ring and offending coil. If significant, the problem can be taken care of by cleaning the affected area and by adding a few layers of new insulation impregnated with insulating resins and insulating paints (see Fig. 4-17).

Fig. 4-17 [S22] An insulated surge-ring supporting the end-winding of a hydrogenerator. Also shown is a resin-soaked and cured winding support cord running inside the coils' knuckles.

S23: Surge-Rings' Support Assembly

The surge-rings (end-winding support rings) restrain the movement of the coils by distributing the forces exerted by one coil onto other coils, and by transmitting them to the frame of the machine. In order to accomplish that, the surge-rings have to remain in sound condition. The inspectors should look for loose parts, bolts and/or nuts, cracked supports made of solid insulation material, greasing bolts, and cracked or loose welding.

As was explained previously, the surge-rings are restrained by the *support assembly* (see Figs. 4-18 and 4-19).

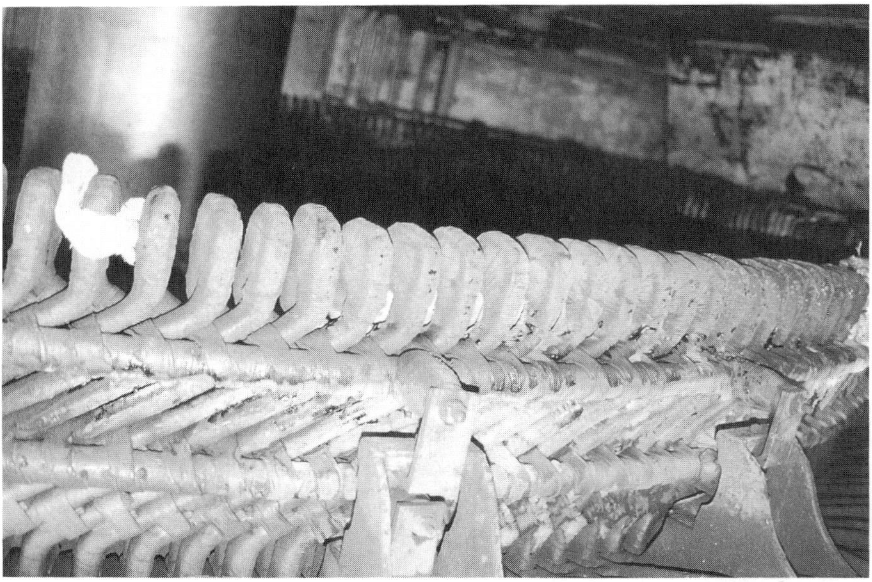

Fig. 4-18 [S23] An end-winding with two surge-rings and their common support assembly.

There is a large variety of support assembly designs. In some cases they are simple steel arms welded both to the surge-rings and to the frame. In such cases each ring has its own set of supporting arms. In larger machines, it is customary to have all surge-rings supported by a more complex assembly.

Experience has shown these to be sturdy and reliable structures. Nevertheless, they should be examined, at least during major inspections, to verify they remain in good condition. As is the case with the surge-rings, the supporting assembly is not readily accessible for visual inspection. It requires some diligence and agility on the part of the inspector. Mirrors can go a long way in facilitating inspection of the support assembly of the surge-rings.

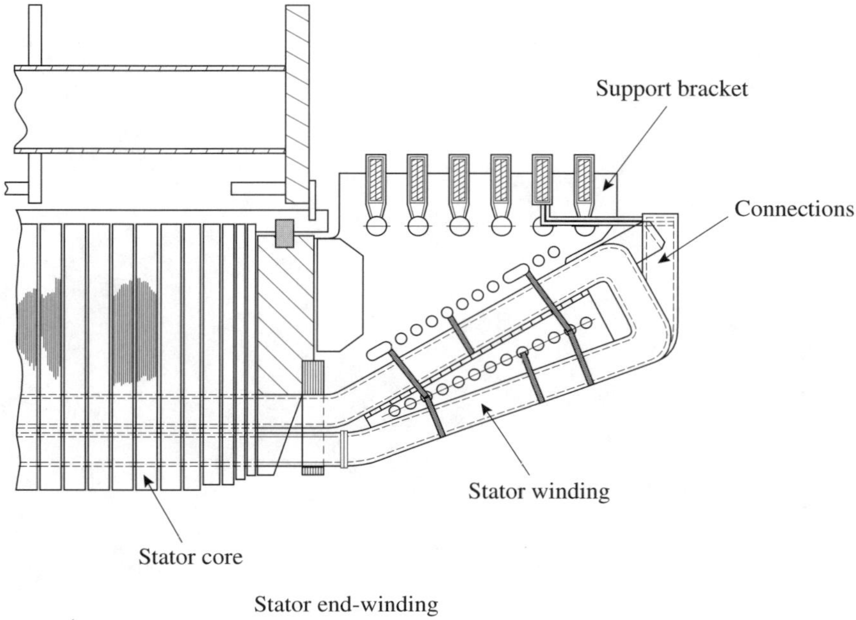

Fig. 4-19 [S23] Schematic representation of a typical end-winding support assembly.

S24: Additional End-Winding Support Hardware

In large synchronous machines, it is common to find other hardware (in addition to simple surge-rings support) that participates in retaining the end-winding structure. This is particularly true in water-cooled machines in which, in addition to the support of the coils themselves, structures also exist to provide mechanical support to the numerous manifolds, water pipes, and other related hardware. One of the critical areas for the integrity of the water system is the condition of the several O-rings connecting different sections of the manifolds (see Figs. 4-20 through 4-22).

It is impractical to describe here the many types of structures found in industry. However, regardless of the construction, the principles for the inspector remain the same.

Look for indications of looseness, fractured parts, loose parts, bolts/nuts, and cracked welding. The treatment for any abnormal condition depends on the availability of spare parts and the type of problem.

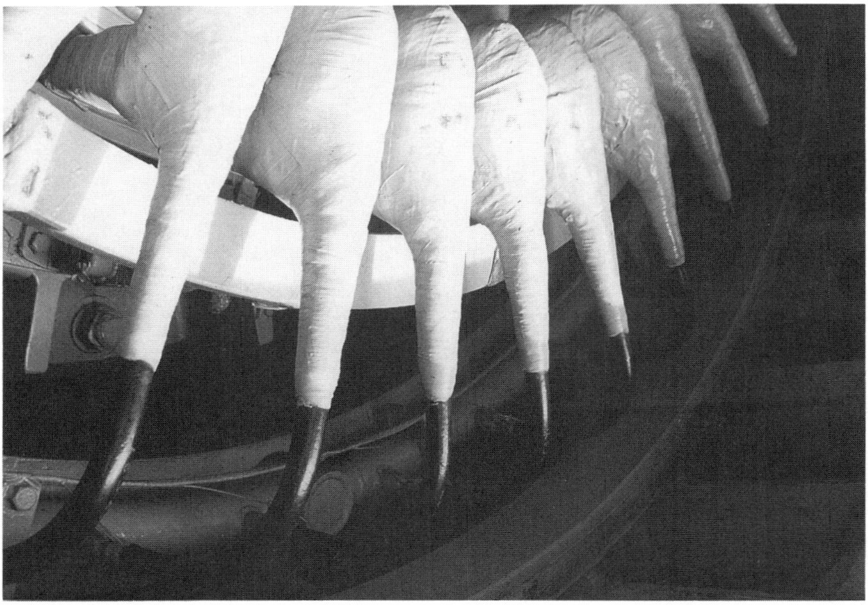

Fig. 4-20 [S24] Portion of an end-winding of the water-cooled stator of a turbogenerator. The water tubes connect to the coils at the coil-knuckle.

Fig. 4-21 [S24] Manifolds and other water-carrying tubing belonging to a 4-pole, 1150-MVA turbine generator.

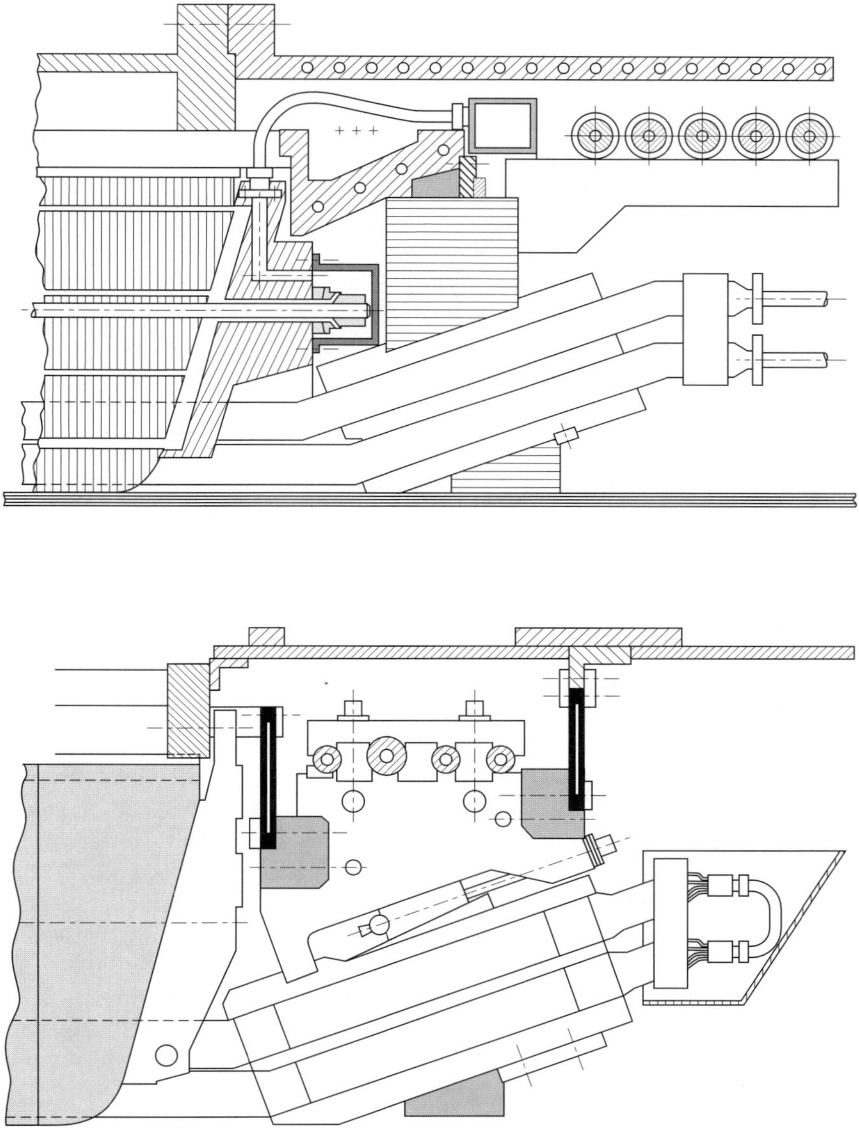

Fig. 4-22 [S24] Schematic representations of the end-windings of water-cooled stators of turbine generators. The lower diagram shows a spring-mounted bracket support system. (Reproduced with permission from ABB.)

S25: RTD and TC Wiring Hardware

Resistance-temperature-detector (RTD) and/or thermocouple (TC) devices are mainly found in windings, cooling gas flow paths, cooling water paths, and bearings. Winding temperature detectors, normally of the RTD type, are located between the coils in inaccessible areas. However, wiring to and from these devices is partially accessible for visual inspection. It should be tightly secured along its path to the coils, frame, and casing. If winding temperature detectors are faulty, they can be replaced during a major inspection or overhaul.

The damaged RTDs are left in place with their wires disconnected or removed, and the new RTDs are taped and glued to the coils, as close to the damaged RTDs as possible. In general, they will be located in the end-winding portion of the coil, after the first bend closer to the iron. Wires broken during the operation of the machine can be identified and repaired.

S26: Asphalt Bleeding/Soft Spots

Sometime during the 1920s and 1930s, the electrical machine manufacturing industry began utilizing asphalt as the bonding agent for the insulation of large synchronous machines. Asphalt was used to bind mica flakes (asphalt micafolium) to form the wall insulation or was used to bond the mica flakes to a tape (asphalt mica tape). In the latter case, the wall insulation was made of a number of layers of tape. The number of layers depended on the tape and the rated voltage of the machine. A final layer of armor tape made of cambric or other materials was commonly used as protection to the cell insulation. In addition to the basic mica and asphalt components, a variety of other elements can be found, such as varnish, asbestos tapes, semiconducting tapes or paints, mylar, and so on, depending on the type of machine, the manufacturer, and the year of production. The coil's fabrication process was completed by a vacuum cycle to remove volatile components and hydraulically pressed. The varnish was applied in a tank or with the tape. Asphalt windings are normally rated as Class B (130°C).

Windings based on asphalt as the binding element were a great improvement over varnished cambric and cells using shellac as the binding material. Asphalt is less prone to voids due to the evaporation of volatiles and is more resistant to water vapor intrusion. Therefore, it is less susceptible to partial discharge and electric treeing. The insulation in the region of the end-windings was a vast improvement over designs made with older insulation systems. Dielectric losses were also reduced. Asphalt allows the ground-wall insulation to become more flexible and less susceptible to cracking and delamination [11]. Asphalt-based coils will expand and fill the slot snugly, reducing (in fact, almost eliminating) abrasion of the wall armor due to vibration of the coil. Also, a tight fit goes a long way in minimizing slot discharges, an important mechanism of ground-wall insulation deterioration. Finally, asphalt insulation possesses better heat-transfer capabilities than older insulation systems.

Asphalt-based coils are included in the group of *thermoplastic* insulation systems. Their manufacture was gradually replaced during the 1950s and 1960s by the so-called *thermosetting* insulation systems.

Given the popularity of the asphalt-bonded windings before the thermosetting systems were introduced, a very large number of machines still operate with those windings. The most common failure mechanisms of these insulation systems are described in this and subsequent numbered items.

In conjunction with the aforementioned advantages, asphalt-based insulation systems are prone to develop a number of problems that are very specific to the class of thermoplastic insulation systems. A major disadvantage of this insulation is its poor thermal resilience. When exposed to high temperatures, the asphalt develops a sharp drop in viscosity and thus tends to migrate along the coil to areas of less pressure. When allowed by a failure in the armor tape, it can even flow out of the coil. Once the thermoplastic element migrates out of a coil section, thermal aging is accelerated due to resulting poorer heat-transfer capabilities. In addition, the excess dryness of the area results in increased partial discharge activity within the wall insulation and in the vicinity of the conductors. If the affected area is large enough, magnetically induced movement of the conductors within the insulation will ensue. This creates internal abrasion with subsequent increase of partial discharge activity. All these mechanisms have the potential to develop into inter-turn and/or turn-to-ground failures.

Asphalt migration might show up as bulging of the insulation in some places. In severe cases, the asphalt will ooze out of the coil, running along it and/or dropping onto other surfaces.

Attempts should always be made to treat areas affected by severe bleeding or migration. The area can be patched with armor tape, rebuilt with epoxy-loaded mica-tape, and any other procedure deemed proper to the type and severity of the damage.

On occasion, migration of the asphalt is masked from view by the armor tape. The coil may appear to be in good condition, without bulging or other visible deformation; however, under the armor tape, a void may be present. Only a spongy or soft yielding to pressure applied by hand on the suspected areas will reveal the presence of a weakened or absent ground-wall insulation. In some cases, the location of the soft or spongy area is not directly accessible by hand. In these cases a probe, preferably nonmetallic, should be used. Care should be taken not to further damage the insulation with the probe.

Given the susceptibility of thermoplastic materials to heat, it is important to avoid any abnormal operation resulting in excessive temperature rise of the windings in machines with this type of insulation. By the same token, when inspecting machines with these insulation systems that were overheated for a relatively prolonged time by overload or insufficient cooling, one should take into account the possibility of severe winding damage.

S27: Tape Separation/Girth Cracking

Note: For background information on the following discussion, see item S26 on page 70.

A problem common to machines with thermoplastic insulation systems (mainly asphalt) is that the normal thermal cycling undergone by the machine, coupled with the snug fit of the coils in their slots, tends to generate cracks within the ground-wall insulation called *tape separation* and *girth cracking.* Although there is some confusion between the two terms (they may mean the same thing in some publications, and different things in others), they can be defined as follows: A tape separation is a separation of the tape covering the wall insulation of the coil due to axial expansion and contraction of the conductors and the opposing forces of the slot applied to the wall insulation. In some cases, only the armor tape is separated. In other instances, the mica-tape comprising the ground-wall insulation is also affected. When this occurs, a few layers may be affected, or in more severe cases, the whole tape will move, creating a "neck" in the coil. Sometimes the ground-wall will separate, forming a sharp crack (see Fig. 4-23).

Fig. 4-23 [S27] Tape separation shown in a coil of a steam-turbine generator.

A girth crack is the necking of the wall insulation, occasionally reaching all the way to the conductor (see Fig. 4-24). Girth cracking occurs both from thermal cyclic stresses in thermoplastic insulation and in insulation rendered excessively dry and brittle by high temperature. After cracks appear, subsequent humidity or other contaminants allow tracking to take effect, with the probable consequence of a short circuit.

Tape separation and girth cracks are commonly found together. They normally appear in machines with core-lengths of about 3 meters (10 feet) and up.

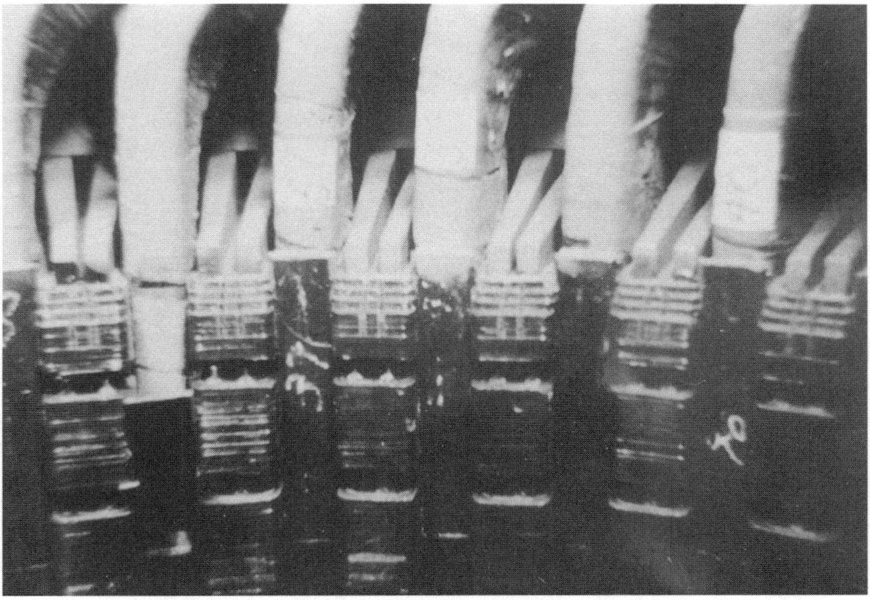

Fig. 4-24 [S27] Tape separation and girth cracks shown in the area where the coils leave the slot. This is the region of the coil most prone to develop this type of problem. Very often, these fissures in the insulation remain hidden underneath the end-wedges and/or fillers. One wedge was removed in this case, clearly showing the girth crack on the coil.

They appear on the end portion of the coils, between the core and the first bend, commonly at about 1 to 2 inches from the core. To a lesser extent, tape separation and girth cracks can be found in the cooling vents (Fig. 4-25).

When substantial tape separation and/or girth cracks are found during an inspection, it is recommended that several wedges be removed to allow inspection of the coil in the slot next to the affected area. Since tape separation and girth cracks can develop under severe conditions in the slot area, removal of several wedges of the suspected coil will indicate if the trouble is localized or if it affects the slot portion of the coil. If the answer is affirmative, more often than not repairs have to be initiated because any degradation of the ground-wall insulation in the slot area has a significant probability of resulting in a short circuit.

Depending on the severity and location of the affected area, the recommended repairs can go from doing nothing to removing a bar. Oftentimes, thin cracks in the insulation are treated with insulating paint. This can help if properly done, with a caveat: The paint can be a carrier for contaminants that will accelerate tracking and possible consequent electric failure. Reference [12] contains an elaborated discussion of the problems herein described, with a range of possible actions aimed at repairing them. The fissure shown in Figure 4-26 could be superficial or serious.

Fig. 4-25 [S27] This coil shows a girth crack in the wall insulation in the area of a cooling vent. The coil was removed from the core, and the tape above the girth was peeled off by the inspector to get a better look at the damaged insulation.

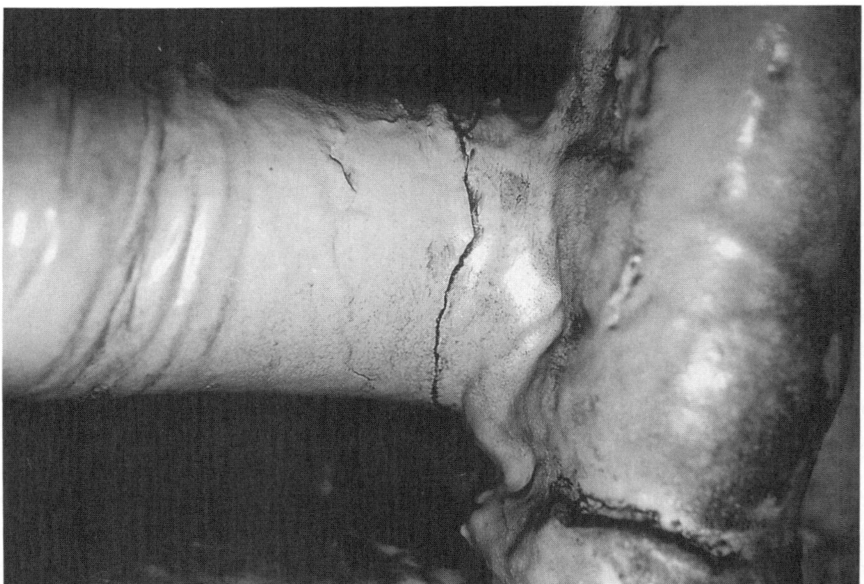

Fig. 4-26 [S27] A fissure on the insulation of a coil near the surge-ring. This could be a superficial crack of the paint or a more serious deep crack.

S28: Insulation Galling/Necking Beyond Slot

Note: For background information on the following discussion, see items S26 and S27.

As explained previously, *necking* or *galling* of the insulation is the result of thermal cycling in thermoplastic insulation systems such as asphalt. Necking is a lack of insulation, which has been cracked and separated. It can also describe the less severe condition of migration of asphalt or other thermoplastic bonding material. Necking is always a sign of a weak point in the ground-wall insulation and should be treated in accordance to its severity and location (see Reference [12] for more hints on repair procedures).

Some stator coils are made in the slot portion with a slot wrapper, while at the end-winding region a tape is used. The interface between the two regions, close to the end of the core, is called a *scarf joint.* Scarf joints represent a weak mechanical point in the structure of the coil, tending to separate under thermal–mechanical cyclic stresses. The indication of a scarf-joint separation is also a necked insulation, or a soft spot. It requires the same type of repairs or attention as previously described for tape separations and girth cracks.

S29: Insulation Bulging into Air Ducts

Note: For background information on the following discussion, see items S26, S27, and S28.

Bulging (balling, puffing) of the insulation right outside the slot and in the cooling vents is an indication of a soft spot, tape separation, girth cracks, or asphalt migration.

When such a situation is encountered, further inspection of the slot area with the aid of a boroscope is recommended. (See Reference [12] for more hints on recommended repair procedures.)

S30: Insulation Condition

To correctly evaluate the condition of a particular winding from test and inspection data, one has to have a minimum of information on the composition of the insulation. Different insulation systems react differently to mechanical, electrical, and thermal aging factors.

Unlike the thermoplastic insulation system discussed previously, *thermosetting* insulation systems, when exposed to elevated temperatures, become dry and brittle. Elevated temperatures may arise from overload conditions, poor cooling (lack of coolant pressure, clogged water pipes, clogged vents, etc.), a damaged core section, negative-sequence currents due to system unbalance, and other causes. Dry and brittle coils can also result from many years of normal operation. For example, machines with old shellac or copal resin binder that have been in operation for almost a century will tend to show this type of condition. It has been the industry's experience that with insulation systems introduced prior to the 1950s, the main mechanism for determining the expected life of the insulation was thermal aging.

In the case of epoxy or polyester binder, severe overload conditions also show up as an external discoloration of the insulation. Dry and brittle windings will sometimes show powder accumulation arising from the movement of the shrunken coil within the slot.

Like other degradation processes afflicting the insulation of machine windings, once a coil becomes too dry and brittle, a positive feedback is established; voids are created within the insulation, further reducing the effectiveness of the heat dissipation from the coil. In addition, internal partial discharge is augmented; looseness and internal movement of the conductors and external looseness of the coil within the slot results in additional mechanisms of degradation taking over, namely, abrasion and slot discharge.

Internal movement of the conductors and the additional partial discharge activity tend to cause failure in these coils in an inter-turn mode. Consequently, a ground fault develops. In the case of single-bar coils, the wall insulation is deteriorated to the extent that a ground fault develops.

Besides visual inspections, the condition of an excessively dry and brittle winding can be assessed with a number of electrical tests, such as partial discharge, insulation power factor, and polarization index.

There is little or no remedial procedure for windings that by aging, wrongful operation, or other reasons become too dry and brittle. Nonetheless, knowing the actual condition of the insulation allows for proper planning of a major rewind and/or better assessment of the risks of continued operation.

S31: Circumferential Bus Insulation

The name *circumferential busses* is given to the solid, circular-shaped phase-connection busses encountered in some multiple-pole synchronous condensers, motors, large turbogenerators, and hydrogenerators. These are supported by studs or other support structures mostly made of steel, bolted or welded to the frame of the machine (see Figs. 4-27 and 4-28).

Circumferential busses are normally separated from the rest of the winding by a relatively large electric clearance. They are commonly insulated. The most common mode of failure in circumferential busses is a breakdown of the insulation adjacent to the metallic supporting studs. The continuous vibration of the heavy copper busses tends to abrade the insulation in this area (see Fig. 4-29).

Continuous movement, together with contamination (the busses are located in close proximity to the rotor fan and are subject to the rapid flow of air or gas containing occasional small amounts of bearing-seal oil, dust, and other contami-

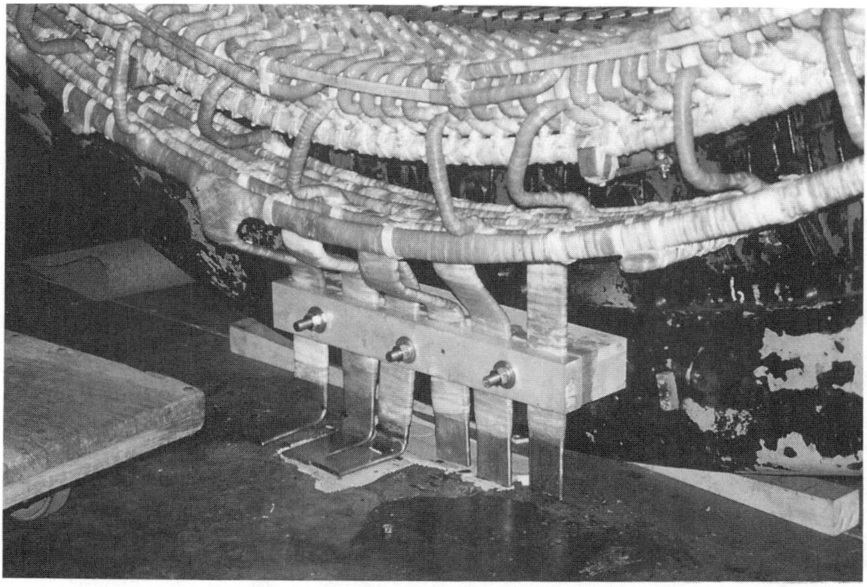

Fig. 4-27 [S31] Circumferential busses at the connection end of a salient-pole machine. Clearly seen are the connections between the circumferential busses, the winding, and the machine's terminals.

Fig. 4-28 [S31] Portion of the circumferential busses in a steam-turbine generator.

Fig. 4-29 [S31] "Greasing" at the region of the support due to movement between the bus and the support structure.

nants), tends to produce tracking over the blocking separating the busses, and may eventually result in phase-to-phase failures. Occasionally, low *megger* readings between the phase under test and the other (grounded) phases can be attributed to the contaminated insulation between circumferential busses.

S32: Corona Activity

Corona activity is defined as the ionization of a gas when exposed to an intense electric field, normally in the vicinity of an electrical conductor. In this form, the definition of corona applies to overhead lines, high-voltage bushings, and other elements producing high concentrations of electric field. In the context of rotating machines, however, the term *corona* is used interchangeably with *partial discharge* and *slot discharge.* This is a high-speed discharge, with a wide range of frequencies (40 kHz to 100 MHz).

Four distinctive types of corona activity can be found within a rotating machine, and are described in the following paragraphs.

1. Corona Activity on the End-Windings. Ionization of the gas at the end-winding region is present in machines operating with line voltages of several thousand volts. The actual inception voltage (i.e., the voltage at which corona is first observed) depends on the specific geometric configuration of the windings and surrounding structures. Different designs render different concentrations of electric fields for the same voltage, thus resulting in varied levels of inception voltages. Corona activity is directly dependent on the actual electric field concentration. Evidently, proper design practices should be geared to minimize high concentrations of electric field in high-voltage machines. Corona activity can normally be found in machines having voltages of 4 kV or higher.

In general, the highest potential differences exist between phase coils (adjacent coils belonging to different phases) and line coils. It is common, therefore, that the telltale signs of this corona activity are concentrated in those areas. The most common signs are white or brownish powder deposits on the coils. In more severe instances, dark burn marks can be found, mainly close to the areas where the coils are at close proximity. Some experience is required to distinguish the corona-originated powder deposits from those originated from fretting of the blocking and ties due to the movement of the coils. Corona-originated powders tend to adhere more tightly to the surfaces and, as mentioned above, tend to be found in areas of high electric field concentration.

Once the machine is in operation, remedial actions will not eliminate the source of corona, as this is design dependent. However, the affected area should be cleaned and the insulation repaired if necessary. These steps will reduce the rate of deterioration of the insulation by secondary phenomena, such as chemical attack of the insulation by accumulated by-products of the corona activity.

It is interesting to note that partial discharge tests are able under certain conditions to identify if the partial discharge activity is inside the coil, or if it is between the slot and the coil. This subject is covered in more detail below. In the case of activity between slot and coil, tightness of the coils and wedges should be checked during the visual inspection of the bore.

2. Internal Partial Discharges. Partial discharge is defined best as an electric discharge occurring between conductors when the breakdown voltage of the surrounding gas is exceeded. When such a discharge is not followed by the establishment of an arc, it is called partial discharge (PD).

PD commonly occurs in voids inside the insulation of the machine. It also occurs between layers of insulation when these are not properly bonded, allowing gaps to remain during the manufacture of the coil or to be created during the operation of the machine. The inception voltage of the discharge depends on the size and shape of the voids, as well as on the gas contained within them. Internal voids created during the manufacture of the coils contain air. Tape separations created during the operation of the machine will be filled with the coolant medium: air in air-cooled machines, hydrogen in hydrogen-cooled units. Machines operating at 4160 V or higher are susceptible to PD activity.

PD activity in voids tends to "eat" away the insulation by abrasion (bombardment of the insulation gas by the acceleration of ions), and chemical reaction with the by-products of the electric discharges in the gas. Occasionally, a void grows into a "tree" due to continued ionic action within the void. Once established, a tree keeps growing at a fast pace due to PD activity and surface discharges within itself. Trees will grow until the ground-wall insulation is weakened to the extent that a full ground fault is developed. Void augmentation or tree creation due to PD activity may also weaken the strand and conductor insulation, giving rise to shorted turns. This phenomenon is most common in the overhang area of the coils, where the manufacturing process does not compress the coils to the extent that it does in the slot section. As a result, more voids are created in the overhang section than in the slot section. A particularly troublesome area is the crossover region of the coils.

Signs of internal PD activity are impossible to detect during a visual inspection of the machine. As will be indicated subsequently, a number of electrical tests are available to assess the extent of PD activity in a particular situation. Nonetheless, visual inspection of the winding and knowledge of the insulation system may provide an indication of the probability of PD activity in the winding.

Bloated or puffy windings, indicating internal looseness, will most probably be subject to high PD activity as well. Dry windings are also susceptible to the existence of layer separation and voids, with accompanying PD activity.

Earlier micaless machines were highly susceptible to failure derived from PD activity. However, practically all machines in operation today include inorganic insulation components such as mica and glass. These components are not seriously affected by partial discharges. On the other hand, the organic materials that make up the bonding structure of the insulation are adversely affected by PD. Old shellac micafolium insulation systems are prone to the formation of voids due to the evaporation of the volatiles in the shellac. The next generation of asphaltic insulation systems went a long way toward reducing voids in the ground-wall insulation. Nevertheless, asphalt-based insulation tends to "swell" or "puff" when a certain temperature limit is exceeded. In addition, weakness of the insulation permits movement between the conductor strands. The consequence of this movement and the generated voids is accelerated PD activity in those spots. All said, thermoplastic (asphaltic) insulation resulted in a great reduction of PD-related faults. Modern epoxy or polyester-based insulation systems offer better bonding, resulting in less internal PD activity than in preceding systems.

Aging due to this type of activity can best be avoided by ensuring no voids are left during the manufacturing process of the coils. Vacuum pressure impregnation (VPI) is such a process. Originally applied to smaller machines, in particular induction motors, VPI windings today can be found in an ever-growing number of synchronous generators. Some European manufacturers use the VPI process in machines with larger than 100 MW ratings—and the upward trend is continuing. However, the repair of coils subjected to the VPI process presents some serious disadvantages.

Other than treating localized areas where a "swell" has occurred, no remedial action exists against a coil or winding afflicted with internal PD activity. The only corrective action, if any, recommended due to the severity of the situation is to schedule a winding replacement in accordance with other planned activities for the unit and to evaluate the risk of continued operation in the present situation.

3. Slot Discharges. This form of partial discharge is the result of the breakdown of the insulating gas between the coil ground-wall insulation and the iron core inside the slot. In alternating current machines, the coil conductors and the opposing slot face together act like a capacitor that is charged and discharged at line frequency. The capacitor "plates" are separated by the ground-wall, strand, and turn insulation, as well as by the insulation of the cooling medium, normally air or hydrogen. Given that the breakdown voltage of air is about 1/100 that of the solid insulation (a somewhat lower ratio for hydrogen), the gas tends to break under the voltage stresses existing in high-voltage machines. The subsequent avalanche of ions abrades the insulation and also attacks it chemically, as already observed previously when discussing internal PD modes of failure. Obviously, coils close to the high-voltage terminals will be subject to higher slot discharges than coils close to the electrical neutral of the machine.

In air-cooled machines, partial discharge is more intensive than in hydrogen-filled machines. Also, discharge in air produces ozone, a very corrosive element. Ozone attacks the organic materials of the wall insulation, accelerating the aging process.

The ozone has a characteristic smell. It is not uncommon for air-filled machines to give away the presence of intensive slot discharge activity by emitting an easily identifiable odor. In these types of machines, it is possible to corroborate the existence of slot discharges by opening inspection plates that allow a view of the slot area of the coils through the core vents, and watching the light emitted during the discharge. Obviously, all lights in the vicinity of the machine have to be turned off for the duration of the inspection.

As with internal PD activity, slot discharge activity is very difficult to identify through a visual inspection of the machine when it is not in operation. Indirect indicators of possible slot discharge activity are loose wedges and coils not snugly fitted in the slots. Asphaltic coils tend to sit tightly in their slots, thus showing diminished tendency to slot discharge activity. Old shellac and modern thermosetting insulation tend to be more susceptible to slot discharges, unless the coils are properly packed and wedged in the slots. Ripple (spring-loaded) fillers go a long way in maintaining a good coil-to-slot pressure.

The solution to the problem of localized slot discharge comes in the form of semiconducting paint or tape, with which high-voltage coils are covered in the slot area. The function of the semiconducting layer is to discharge all surface charge through the contact points between the coil and the walls of the slot. However, enough points of contact should be established and maintained between the semiconducting paint and the slot to maintain the integrity of the paint or tape. If continuous movement of the coil within the slot occurs, the fretting and burning of the few contact points deteriorate the semiconductor paint or tape to the extent that they cease to protect the ground-wall insulation from the increased slot discharges (see Figs. 4-30 and 4-31). Eventual failure of the insulation follows. To maintain the integrity of the semiconductor layer, the coils have to sit snugly inside the slot.

Fig. 4-30 [S32] Coils removed from a salient-pole generator. The coils show severe loss of their semiconducting paint due to the coils not being properly supported in the slots.

4. Surface Discharges. Surface discharges are intermittent partial breakdowns of the surface insulation due to high electric fields. The effect of these surface discharges is damage to the organic surface layer of the wall insulation. These discharges, while less damaging than the three types already discussed, do contribute to the general degradation of the insulation.

Surface discharges tend to be concentrated on the overhangs of the windings, in the immediate vicinity of the core. Grading paint, a feature introduced in high-voltage machines to eliminate high concentrations of electric fields at the end-region of the core, also controls the surface discharges generated in this region.

Fig. 4-31 [S32] Close view of one of the affected areas shown in Figure 4-30.

As stated previously, some mechanisms of PD or corona activity cannot be readily identified during visual inspection of the machine. Several tests have been designed to evaluate the presence and intensity of PD activity and the resulting damage to the coils.

All tests designed for the detection of corona or partial discharge activity have to be performed with the machine energized. Some of those (e.g., requiring the use of a hand-held probe) require the rotor to be removed from the bore.

Some tests, such as the polarization index (PI), dielectric absorption, power factor (PF), power factor tip-up (PF tip-up), and radio frequency test (RIV), are general insulation tests that also provide information on PD activity and the damage accumulated in the windings due to PD.

Other tests are specifically designed for the detection of partial discharge; for instance, the Integrated Discharge Energy Measurement (Westinghouse), the embedded stator slot coupler (SSC), partial discharge measurements, electrostatic probe tests, and ozone meters.

Figure 4-32 shows the installation of an embedded SSC. (Reference [10] contains a very good description of the tests. References [13]–[21] contain additional information on corona/PD activity.)

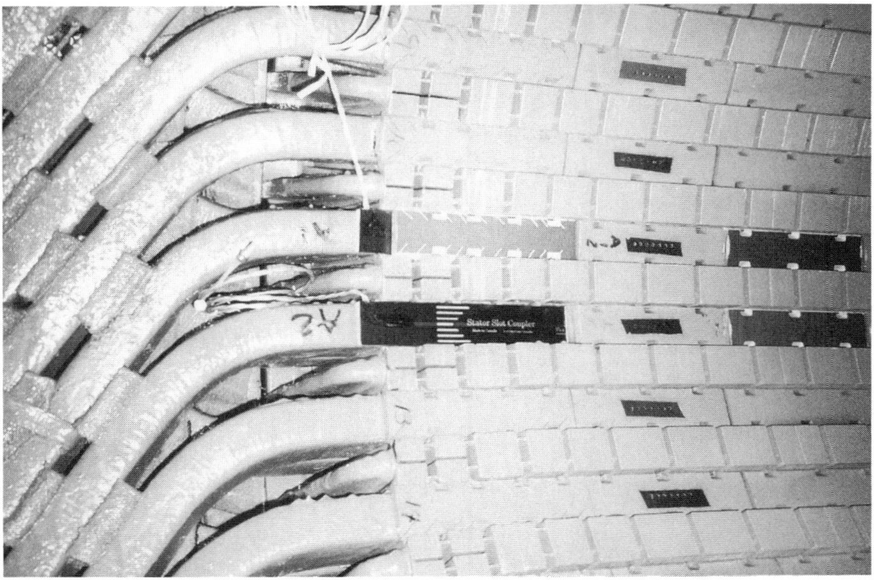

Fig. 4-32 [S32] The embedded stator slot coupler (SSC) being installed under the wedges of a large turbine generator.

S33: Wedges Condition

One important item for inspection are the stator wedges. The wedges are one of the main elements controlling the tightness of the coils in the slots. Maintaining a positive pressure on the coils reduces their movement within the slots, thus minimizing loss of semiconducting coating and wall insulation.

The common way to inspect wedge conditions is to tap on one side of the wedge with a small hammer and sense the amount of movement with the other hand touching the other side of the wedge (see Fig. 4-33). Given the response, the wedge condition can be classified either as tight, loose, or hollow. A loose wedge is a wedge that responds to the tapping with movement. A hollow wedge is one that doesn't move but sounds like it is not pressing against the coil. It takes some experience to differentiate between hollow and tight wedges; however, this experience can be readily acquired.

A hollow wedge indicates a clearance exists in the radial direction between the slot and the coil. This clearance could be due to poor packing during installation of the coils, or coil shrinkage. Hollow wedges can be found more often in thermosetting coils than in thermoelastic coils, which tend to fill the slot.

A substantial number of hollow or loose wedges in a row indicate a coil is loose and will probably tend to move within the slot during operation. This is an undesirable condition. If enough wedges are found to be loose or hollow, a wedge

Fig. 4-33 [S33] Movement of loose wedges in the presence of oil creates the telltale greaselike deposits shown in this photograph.

survey might be desirable. During a wedge survey, all wedges are tapped and the response recorded in a form similar to Form 9 (included in Chapter 3). The wedges are numbered by their location both by slot number and position within the slot. If a number of contiguous wedges are not acceptable (hollow or loose), then the slot might be re-wedged. Each operator or inspector uses different criteria to decide on the threshold number of unsatisfactory wedges beyond which a "re-wedge" is performed. A commonly used number is 25% of the total number of wedges. End-wedges should always be secured if found loose; application of epoxies and RTV is one common technique of securing these wedges. The decision to re-wedge, perform partial repairs, or continue operation "as is" depends also on the importance of the machine to the system, repairs costs, and so forth.

A wedge survey performed with the hammer method can be very tedious and time-consuming. New hand-held "tapping" instruments are available. These instruments allow the survey to be conducted in a fraction of the time and with more consistency of readings, in particular when different inspection personnel are involved. In many cases, people tend to interpret differently the response obtained by the "hammer" method. The automatic tapping instrument provides an important measure of consistency between results obtained by various inspectors.

In newly designed machines, or recently re-wedged machines, a "ripple" filler under the wedge tends to reduce wedge looseness. In addition, these modern wedges have, on many occasions, been manufactured with small orifices through which a probe can be inserted to measure the tightness of the coil against the wedge. Some organizations have recently introduced to the market a robotic method of tapping and/or probing the condition of the wedges in large synchronous machines without the need for removing the rotor from the bore.

One way to identify loose wedges is by the resulting greaselike or powder deposits along the wedge edges (due to the continuous movement of the wedges). The powder deposits tend to be yellowish or reddish, due to iron oxide resulting from the wedge-core fretting action (see Figs. 4-34 and 4-35).

If only a few wedges are found loose, a common remedy is to apply epoxy or other resins.

Fig. 4-34 [S33] Close-up view of a loose end-wedge and the "greasing" (along the bottom edge of the wedge).

Although it was stated above that a value of 25% of loose and/or hollow wedges can be used as a criterion for re-wedging the machine, it has to be recognized that the distribution of those loose/hollow wedges may call for different actions. For instance, a machine may have a relatively small percentage of loose/hollow wedges, but have them concentrated in one particular area of the stator. This high concentration of unsatisfactory wedges may require partially re-wedging the stator.

In another example, 15 to 20% loose/hollow wedges evenly distributed in one slot may provide enough support to the coil; on the other hand, the same 15 to 20% loose/hollow wedges all concentrated in one side of the slot will require re-wedging that slot.

Fig. 4-35 [S33] Tapping wedges inside the bore of a 475-MVA, 4-pole generator.

S34: Wedges Slipping Out

It is a common occurrence for the end-wedges (wedges on both ends of the core) to become loose during operation of the machine. In older machine designs, particularly hydrogenerators or salient-pole condensers and motors, the end-wedges tend to slip out of the core.

This condition can be readily detected by inspection with the naked eye or with the aid of mirrors or a boroscope when the rotor is in the bore. A commonly used fix comprises the application of a thick resin, epoxy, or RTV type of material. A substantial number of end-wedges moving out of the slot may indicate an overall loose wedge condition, and might warrant a wedge survey (see item S33).

Newly designed machines and most turbine generators have locking designs that do not allow the outward movement of the end-wedges.

S35: Fillers Slipping Out

Another indication of loose coils, at least in the radial direction, is the movement out of the slot of the top and bottom fillers. This phenomenon does not exist in windings processed with VPI. In all other types of windings, it is not uncommon to observe during extended inspections the movement of top, bottom, and side fillers out of the slots (see Fig. 4-36).

Fig. 4-36 [S35] Side-coil fillers belonging to a salient-pole synchronous machine, slipping out of the end of the core approximately one inch.

Normally, the fillers are driven back (if possible) or broken at the end of the core. In both cases, they are secured with resin, epoxy, or RTV materials.

As with the movement of end-wedges, large numbers of fillers slipping out of the core by several inches may indicate a loose winding condition. However, filler movement can also be the result of elongation and contraction of the coils due to thermal cycles, even in tight coils. A partial wedge survey might provide the required information on the snugness of the coils inside the slots.

S36: Bars Bottomed in Slot

During the visual inspection of the stator winding, it is recommended to verify that the coils are seated tightly on the bottom of the slots. A mirror permits inspection around the end-windings and observation of the end core area and the bottom of the coils. Coils not bottomed indicate a loose-coil condition with all the problems and consequences detailed above in item S33.

S37: Laminations Bent/Broken in Bore

Core laminations are often damaged during the removal of the rotor (see Figs. 4-37 and 4-38). When this occurs, it is not uncommon that they become partially

Fig. 4-37 [S37] A number of bent laminations from the first packet (closest to the edge of the bore). They were bent during removal of the rotor. The machine is a 2-pole turbine generator.

Fig. 4-38 [S37] Same machine as in Figure 4-37. Bent laminations shown at different packets.

short-circuited. If left in this condition, they may reach excessive temperatures during operation of the machine, resulting in damage to the insulation between laminations. It is therefore recommended that laminations or groups of laminations bent during the removal of the rotor be straightened carefully, and their tops separated and impregnated with resin or epoxy so they do not remain short-circuited.

It is obvious that care should be taken during replacement of the rotor into the bore to avoid damage to the laminations. Any damaged laminations will remain damaged until the machine is disassembled again, commonly several years later.

As already stated in other sections of this book, it is most important to identify laminations that are broken or in the process of breaking up. Any broken pieces of lamination will get loose and most probably will damage the insulation of the coils. Small pieces of lamination can "drill" into the coil insulation by the continuous action of the double-line-frequency magnetic forces that the field of the machine exerts over these loose pieces of metal. Some people call these loose small pieces of metal "magnetic termites." Bent laminations are suspected of having experienced excessive metal fatigue. Therefore, the inspector should always assess the source of the bending and the condition of the lamination.

S38: Laminations Bulging into Air Ducts

On rare occasions, inspection of a generator's bore reveals that the core laminations are bulging into the air ducts or vent areas, as well as bending away from the core in both ends (see Figs. 4-39 and 4-7). These problems normally occur as a result of weakened lamination support, combined with vibrations and coil movement.

In some designs, the duct-spacers are curved toward one corner of the tooth to direct the flow of the cooling gas. In this design, laminations may bulge in one corner of the tooth.

Laminations with weakened support at both ends of the core are subjected to lateral "flapping" by the magnetic forces induced by the twice-supply-frequency axial flux present at those regions of the machine. Eventually, these laminations tend to crack and break due to metal fatigue (see Fig. 4-7) if not glued together during an overhaul.

In cases of extreme bulging, the only solution might be a new core. In milder conditions, the solution might take the form of restacking the core, or restacking part of the existing laminations together with new ones.

Taking no action will limit the output of the machine due to overheating arising from the restriction on the flow of air (or hydrogen), as well as other heating problems associated with the extra losses generated in the core.

Fig. 4-39 [S38] Lamination bulging into a vent duct in a steam-turbine generator. In this case, the offset position of the duct spacer provides weak support on one side of the laminations' packet. If this condition is left untreated, continuous vibration of the loose lamination(s) may cause them to break.

Migration of Duct Spacer Assemblies. During the operation of large steam-turbine generators, the migration of duct-spacer assemblies has been recorded. These assemblies contain the I-shaped separators that maintain the spacing between packets of laminations required by the cooling gas to circulate radially within the core.

Migration of dust-spacer assemblies is more common in older hydrogenerators, with cores made up with a large number of lamination segments.

When migration of the duct-spacer assemblies has occurred, this has been toward the center of the machine; i.e., toward the inner diameter of the bore. Under these circumstances, the bottom of the slot of the lamination damages the bottom coil's ground insulation (see Fig. 4-40). The result can be a ground fault, or worse still, multiple ground faults from coils at different phases, resulting in between-phase short circuits across the laminations. In this instance, the cost of repairs in time, money, and material are generally very substantial.

This migration condition can be safely diagnosed only by inspection of the bore with the rotor removed. Any migration of the duct-spacer assembly should be promptly discussed with the machine manufacturer.

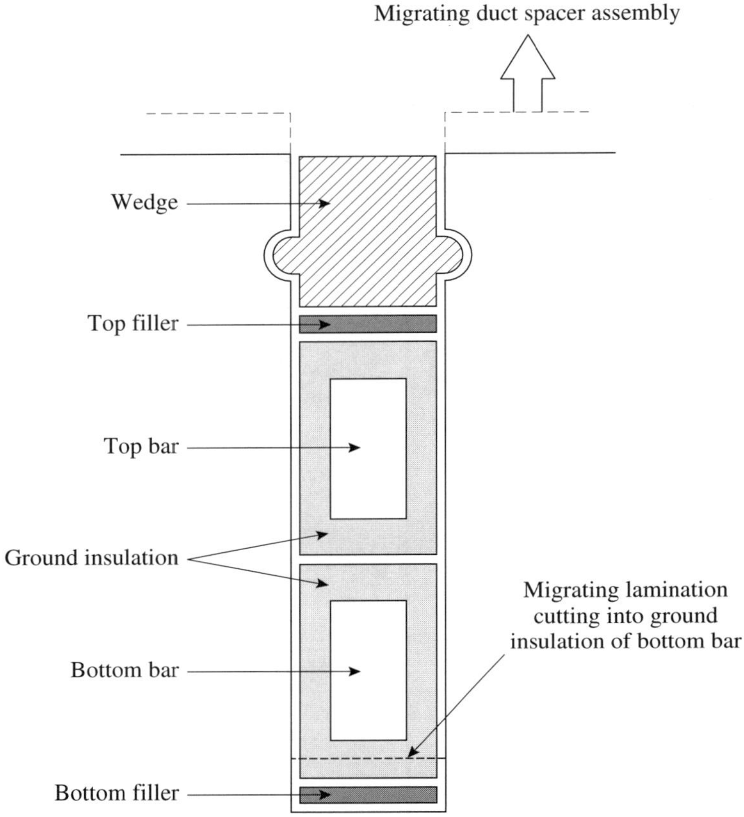

Fig. 4-40 [S38] Cross section of slot showing migration of duct-spacer assembly.

S39: Terminal Box Current Transformer Condition

As part of any major inspection of a large synchronous machine, the condition of the main current transformers (CTs) should be evaluated. In large generators, the CTs are often located underneath the machine, just below the terminal box or bushing well. The CTs, which normally come as two or three per phase, are placed around the phase and neutral iso-phases where these leave the bushing well. In smaller, typically air-cooled machines, the CTs may be located inside the terminal box.

The CTs are normally mounted on insulated bolts. Connection to the CTs is carried out by grounded metal piping containing the connection leads.

Large machine CTs are commonly manufactured with their windings and core encased in resin-type material. On top of this, an aluminum envelope is typically added.

Internal faults of the current transformer winding that result in substantial heat being generated tend to show up as cracks in the exterior of the aluminum shell and/or as flow of the encasing resin out of the shell.

Signs of overheating, such as leaking resins or discoloration, provide a strong reason for further investigation. The surfaces of the CTs should be cleaned to minimize any possibility of flashover.

Like any other current-carrying element of the power plant, CTs can be scanned with an infrared camera while the machine is in operation to check for abnormal temperatures.

S40: Bushing-Well Insulators and H_2 Sealant Condition

In large synchronous machines, the connection of the windings to the armature leads is carried out inside a large terminal box (bushing well) underneath the machine. Often, foreign material such as oil, loose bolts, washers, fragments of insulation, et cetera, become lodged in this compartment. Hence, it is important that the terminal box is opened and inspected at least during major inspections. In addition, the condition of the bushings and stand-off insulators should be inspected (see items S05 through S07). The resin tape on the terminations should also be inspected for integrity.

These bushing wells normally have small amounts of a viscous sealant oil in the bottom of the well. The purpose of this sealant is to eliminate the leak of gas through the mounting flanges of the high-voltage bushings. The condition of the sealant should be evaluated during an inspection inside the bushing well. Dry or contaminated sealant oil should be removed, and new material should be brought in.

S41: End-Winding Expansion-Bearing-Bolts Condition (Round-Rotor Machines)

In some large turbogenerator designs, the end-winding support assembly is allowed to move in the axial direction to accommodate the thermal expansion and contraction of the winding. The movement is allowed by having the whole end-winding support system mounted on so-called *expansion-bearing-bolts.* Continuous vibration may cause fretting and eventually breaking of the bolts assembly (see Figs. 4-41 and 4-42). This situation may encroach upon the free movement of the end-winding support system, giving rise to uneven stress. The stress may damage the integrity of the end-ring support system. Also, some indications exist

Fig. 4-41 [S41] A winding support bearing-bolt. "Grease" deposits shown on the bolt are an indication of looseness.

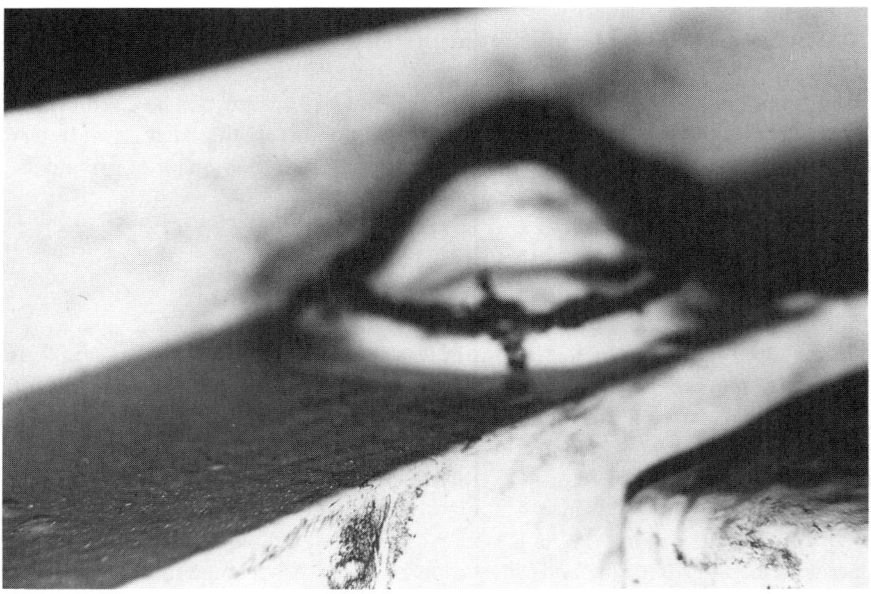

Fig. 4-42 [S41] "Grease" deposits at the head of the bearing-bolt of Figure 4-31, indicating a loose bolt.

of abnormal vibration caused by faulty expansion-bearing-bolts, providing insufficient support to the end-windings. Therefore, the expansion-bearing-bolts, when present, should undergo close examination during a major inspection.

REFERENCES

[1] ANSI/IEEE Std 67-1972, "IEEE Guide for Operation and Maintenance of Turbine Generators," Item 8.6.2, p. 34.

[2] H. R. Tomlinson, "Inter-laminar Insulation Test for Synchronous Machine Stators," *AIEE Transactions,* Vol. 71, Part III, August 1952, pp. 676–677.

[3] *EPRI Power Plant Electrical Reference Series,* Vol. 16, Items 5.2.4 and 5.2.5, p. 5-42, and Item 6.5.4, p. 6-22, 1991.

[4] ANSI/IEEE Std 56-1977, "IEEE Guide for Insulation Maintenance of Large Alternating-Current Rotating Machinery (10,000 kVA and Larger)," Item 6.4.1, p. 10.

[5] *EPRI Power Plant Electrical Reference Series,* Vol. 1, pp. 1-11, 1-37, 1-47, 1991.

[6] GEI-37081: "Instructions on Hydrogen-Cooled Turbine Generators, Mechanical and Electrical Features," General Electric.

[7] J. Boyd and H. N. Kaufman, "The Causes and Control of Electrical Currents in Bearings," *Lubrication Engineering,* January 1959, pp. 28–35.

[8] G. W. Buckley and R. J. Corkings, "The Importance of Grounding Brushes to the Safe Operation of Large Turbine Generators," *IEEE Transactions on Energy Conversion,* September 1988, pp. 607–612.

[9] Jim Sparks, "What to Do about Shaft Currents," Plant Operation and Maintenance Section, *Power Generation and Transmission,* p. 114.

[10] The Doble Engineering Company, "Rotating Machinery Insulation—Test Guide," 1985.

[11] *EPRI Power Plant Electrical Reference Series,* Vol. 16, Item 2.1.1, pp. 2-3 and 2-4; Item 2.3.2.7, p. 2-9; Items 3.1.2.1–3, pp. 3-18, 3-20 to 3-21, 1991.

[12] C. A. Duke et al., "Visual Inspection of Rotating Machinery for Electrical Difficulties," Conference of Doble Clients on Rotating Machinery.

[13] E. H. Povey, "Corona Measurements by the RIV Method," Conference of Doble Clients on Rotating Machinery, 3-201, 1958.

[14] W. A. Patterson, "Testing for Corona in Generator Stator Windings," Conference of Doble Clients on Rotating Machinery, 7-601, 1967.

[15] W. A. Rey, "Increased Deterioration of Generator Insulation by Corona Action," Conference of Doble Clients on Rotating Machinery, 7-101, 1951.

[16] C. A. Duke, "Experience with Slot-Discharge Testing on Generators," Conference of Doble Clients on Rotating Machinery, 7-101, 1958.

[17] IEEE Std 226, "Recommended Practice for Measurement of Power-Factor Tip-Up of Rotating Machinery Stator Coil Insulation."

[18] T. W. Dakin, "The Relation of Capacitance Increase with High Voltages to Internal Electric Discharges and Discharging Void Volume," *AIEE Power Apparatus and Systems,* Vol. 78, pp. 790–795.

[19] T. W. Dakin, "A Capacitance Bridge Method for Measuring Integrated Corona-Charge Transfer and Power Loss per Cycle," *AIEE Power Apparatus and Systems,* Vol. 79, pp. 648–653.

[20] W. McDermid, "Review of the Application of the Electromagnetic Probe Method for the Detection of Partial Discharge Activity in Stator Windings," *Proc. of the CEA International Symposium on Generator Insulation Tests,* Toronto, 1980.

[21] S. R. Campbell, G. C. Stone, et al., "Practical On-Line Partial Discharge Tests for Turbine Generators and Motors," *IEEE Transactions on Energy Conversion,* Vol. 9, No. 2, 1994.

ADDITIONAL READING

EPRI Report EL-3564-SR, "Workshop Proceedings: Generator Monitoring and Surveillance," August 1984.

EPRI Report NP-902, "On-Line Monitoring and Diagnostic Systems for Generators," September 1979.

5

Description of Rotor Items

Chapter 5 describes each item on Form 5, the Rotor Inspection form. Rotor inspection encompasses both round-rotor and salient-pole machines. If an item is found only on a round-rotor machine, the item is marked (RR) both on the form and in its corresponding heading in this chapter. Similarly, if an item is found only on a salient-pole machine, the item is marked (SP) both on the form and in the corresponding heading in this chapter.

R01: Rotor Cleanliness

Cleanliness, or a measure of cleanliness, is important not only to the proper operation of the machine but also to provide the inspector or maintenance crew with clues on the overall condition of the machine.

For instance, a solid rotor exhibiting numerous deposits of copper powder indicates excessive movement of the DC field coils when on turning gear.[1] Excessive copper dust should alert the inspector to the possibility of the existence or the development of shorted turns and/or ground faults. Under those circumstances, the inspector may decide it will be prudent to carry out one of the available shorted-turn tests. In addition, the machine's owner may decide to change the blocking, insulation, and so on. Ground faults can easily be detected with a *megger* (generic name for an instrument that measures insulation resistance).

Copper dust, iron dust, or any other telltale material may be concealed in a mixture of oil and dirt. In some instances, chemical analysis of a sample is required to assess the true amount of copper and/or iron dust.

Dirt can mask cracks on the surfaces of critical components such as wedges, fan hubs and blades, retaining rings, the forging itself, the amortisseur bars' connection to the short-circuiting rings in salient-pole machines, etc. Therefore, critical areas should be cleaned to the full extent of their accessibility. However, it is important that the inspection be carried out both before and after cleaning. Before-cleaning inspections reveal a wealth of information on the rotor or machine condition and operation history, while inspection after cleaning allows the evaluation of the conditions of rotor components.

Besides masking important areas from the inspector, heavy dirt deposits can adversely affect the flow of the cooling gas or air, in effect de-rating the machine.

R02: Retaining Rings' Visual Appearance (RR)

The retaining rings are the most critical component in the rotor and normally the most highly stressed rotor component. These rings are critical in the sense that their mechanical failure always has catastrophic consequences on the physical integrity of the machine. Therefore, retaining rings deserve the utmost attention from the inspection team.

Manufacturers of large synchronous machines usually issue periodic informative bulletins. In each bulletin, the attention of the users of a particular type of machine is brought to problems experienced by other users of machines with similar components. Recommended remedial actions necessary to avoid or minimize risks of similar problems are presented. In the case of retaining rings, a substantial amount of information has been disseminated by the original equipment manufacturers (OEMs), clients, and organizations such as the Electric Power Research Institute (EPRI). In general, the manufacturers' recommendations for the protection of the rings should be followed.

Periodic inspection of the retaining rings is highly recommended. The actual length of time between inspections depends on the mode of operation of the machine, which includes how many hours it operates and, in particular, whether it is in continuous operation or experiences many start–stop cycles. The successive thermal cycles accompanying a machine undertaking many start–stop operations tend to subject the retaining rings to accelerated metal fatigue.

A major cause for advancing the inspection schedule of the retaining rings is moisture penetration on a machine equipped with the older 18-5 type of ring (18% manganese-5% chromium alloy). The modes of moisture-induced failure of the 18-5 rings are discussed in Chapter 4 under item S13.

Other reasons for performing at least a partial inspection of the retaining rings could be excessive asynchronous "motoring" or generation without the DC field active. Under these conditions, the machine acts as an induction motor or generator. This can be damaging to rotor components such as wedges and retaining rings due to the flow of induced currents in those components. Negative-sequence currents also tend to flow in the forging, wedges, and retaining rings.

An area prone to damage from arcs is the contact area between the rings and the adjacent rotor wedges. This area should be checked for electric pitting or discoloration, indicating the flow of currents. If these symptoms are found, remedial action should be taken. To minimize the risk of current flow between the wedges and the retaining rings, it is good practice to move the wedges away from the rings during overhauls when the rotor is out of the bore.

Through visual inspection of the retaining rings, a preliminary assessment can be made as to the need to carry out specific tests such as nondestructive examinations (NDEs) of the several types available: eddy current, acoustic, die penetrant, as well as hardness tests, and so forth. If the visual inspection of the surface reveals oxidation traces, pitting, or other warning anomalies, inspection of the inner side of the retaining rings might be advisable (see Fig. 5-1). This entails removing the rings, a costly and somewhat risky operation (see Fig. 5-2).

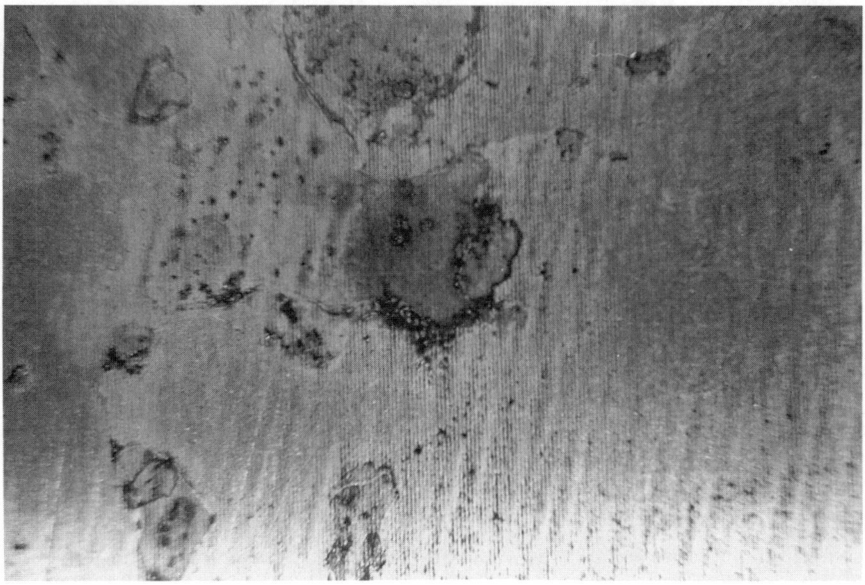

Fig. 5-1 [R02] Portion of a retaining ring showing rust deposits where the protective paint has peeled off. Very often the rust is superficial and easily removed. In any instance, the presence of rust should alert the inspector to the possible existence of water-originated pitting.

It goes without saying that for visual inspections to be effective, the paint (if present), oil, and dirt on the surface of the rings have to be removed. Unfortunately, while the paint is removed, the polishing action may conceal otherwise visible trouble spots. Some manufacturers recommend not to paint the retaining rings. This depends on the steel type and cooling medium, as well as the manu-

Fig. 5-2 [R02] The retaining ring of a turbine generator's rotor is being prepared for its removal by electrical heating. Other commonly used techniques are induction heating and gas flame heating.

facturer's own experience and preferences. Some vendors offer technologies that can perform nondestructive examinations without the need to remove the paint.

During major inspections, NDEs and hardness tests are always recommended. Commonly these tests demand the removal of the rotor. Recently, however, some service providers have been able to perform a number of nondestructive tests on retaining rings without removing the rotor from the bore.

Several rotor designs have holes drilled in the body of the retaining rings (Figs. 5-3 and 5-4). These allow the flow of gas for cooling purposes. Ring ventilation holes tend to be areas of mechanical stress concentration. The tangential expansion forces cause the areas of maximum stress at the holes to be at the inner diameter (ID) of the ring in the axial direction; i.e., facing the ends of the rotor. Some people highly recommend that drilled retaining rings have eddy-current NDE performed on all holes. Other areas of stress concentration that should be inspected closely are tapped holes for the locking arrangement on the circumferential ring-keys.

It is extremely important to maintain a dry and clean environment around the rotor not only during the operation of the machine but also during the overhaul and inspection activities. Of particular importance is keeping humidity out of the retaining rings made of 18-5 alloy steel. Corrosion occurring when the rotor is removed from the bore can lead to failure of the ring during future operation.

Fig. 5-3 [R02] The retaining ring of a 2-pole turbine generator with vent holes.

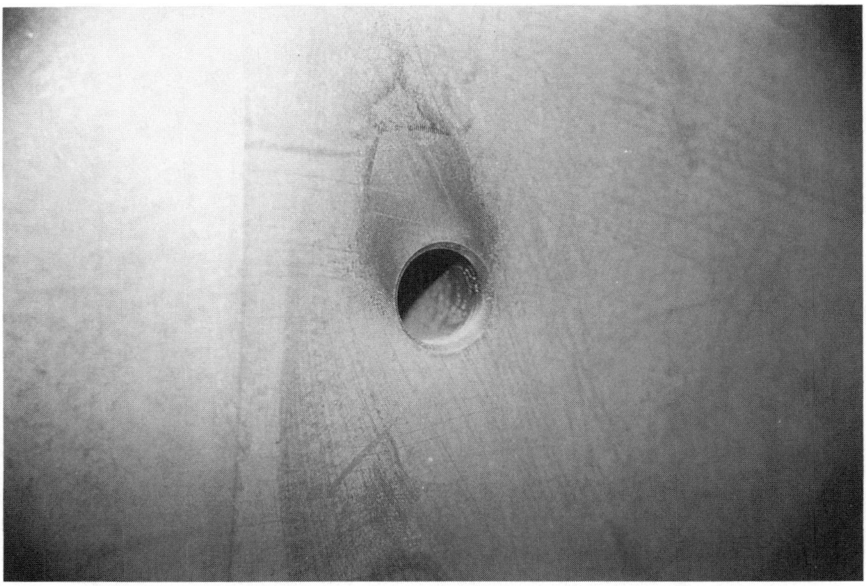

Fig. 5-4 [R02] A close view of a vent hole from the retaining ring in Figure 5-3. The shadow is the result of deposits and erosion by the continuous high-speed flow of gas leaving the vent in the direction of the shade.

EPRI's publication EL/EM-5117-SR has valuable information on guidelines for the evaluation of the condition of retaining rings, as well as a list of vendors offering specific inspection tests [1].

There are two basic designs for retaining rings: the *spindle-mounted* design and the *body-mounted* design. Each design has characteristic weak points demanding close examination during an inspection. Figure 5-5 shows the basic retaining ring designs and their weak points.

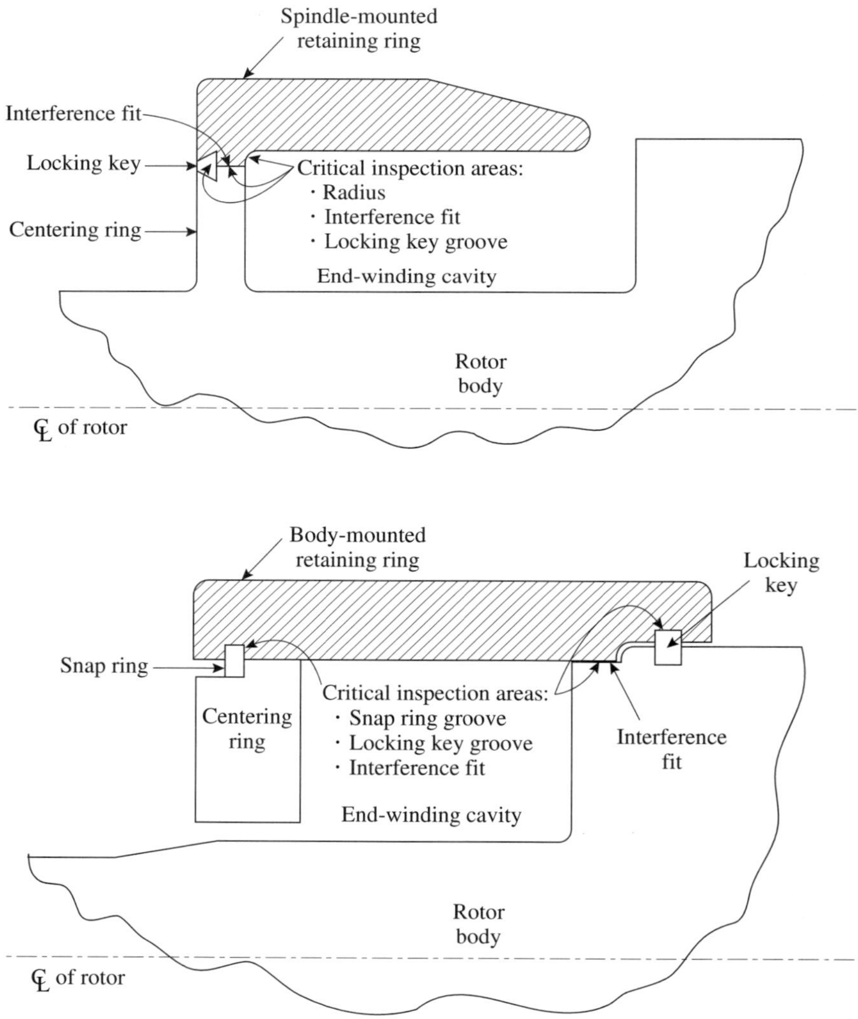

Fig. 5-5 [R02] Schematic representation of typical spindle-mounted and body-mounted retaining rings.

It is important to note that when removing a body-mounted retaining ring, the inspector should look for the so-called *tooth-top cracking* phenomenon. This alludes to when the top part of a rotor tooth directly under the shrunk retaining ring cracks or breaks due to the mechanical stress to which it is subjected. In severe cases, parts of the tooth come loose during removal of the retaining ring. Some designs are more prone to this phenomenon than others. Whenever a body-mounted retaining ring is removed, however, the inspector should consider looking for cracked rotor teeth.

R03: Centering Rings' Visual Appearance (RR)

Retaining rings are shrunk onto centering rings. The interference surfaces include arrangements for the presence of locking keys and locking-key grooves. These are areas of concentrated stress. They should be inspected for stress-related cracks and other anomalies.

R04: Fan-Rings' Visual Appearance

In 2- or 4-pole machines, the fan blades are attached to fan-rings shrunk onto the shaft (Figs. 5-6 and 5-7). In low-speed machines, the fans are occasionally at-

Fig. 5-6 [R04] The fan ring and fan blades of a 2-pole cylindrical rotor. In this design the diameters of the fan ring and blades are larger than the diameter of the bore. Thus, the rotor can only be removed in the direction of the fan unless the fan is disassembled prior to removal of the rotor. Compare this design with the one shown in Figure 5-7.

Fig. 5-7 [R04] A 2-pole cylindrical rotor with its axial fan having an overall diameter smaller than the diameter of the retaining rings.

Fig. 5-8 [R06] Salient-pole synchronous machine with radial fans bolted to the rotor's pole-support structure.

tached to a large ring or circumferential plate, as shown in Figure 5-8. In either case, interference-fit stresses and/or centrifugal forces induce mechanical stresses in those elements. It is therefore important during a major inspection to assess their mechanical integrity.

R05: Fretting/Movement at Rings' Interference-Fit Surfaces

Look for fretting or other movement signs between the contact surfaces of shrink-fit components. Signs of fretting or movement could indicate excessive heating of the shrunk member or abnormal forces experienced during the operation of the machine. Examples are cracks of the shrunk member, insufficient interference fit, and so on. Given the serious consequences of failure of any rotating element, prompt attention should be given to any such sign of distress.

R06: Fan Blades Condition

Two-pole and 4-pole machines tend to have axial fans on their rotors (Figs. 5-6 and 5-7). Machines with a larger number of poles tend to exhibit radial fans (Figs. 5-8 and 5-9). In either case, the fan blades might be an integral part of the

Fig. 5-9 [R06] Arrangement similar to Figure 5-8, but in a smaller machine.

fan hub; i.e., part of the casting, either welded or bolted to the fan-hub or plate or to the body of the rotor itself (see Figs. 5-10 and 5-11).

In any of the many versions of fan-blade attachment, it is very important to examine the condition of the blades and the soundness of their attachment. Cracks at the root of axial fan blades are not uncommon. If bolted, attention should be given to snugness and nut-lock condition.

R07: Bearing Journals Condition

Depending on whether the machine is vertical (mostly hydro) or horizontal, many kinds of bearings can be found. These will almost always be of the friction type for large machines. The diversity of bearings makes it impossible to describe them all here. The following comments apply to any type of bearing.

Inspection of the bearing journals is part of any extensive machine inspection. It is, of course, combined with inspections of the babbitt metal in horizontal machines and pads in vertical ones, as well as the inspection of the oil-baffle labyrinth, and measurement of oil-seal ring clearances and bearing clearances.

Inspection of the used oil may shed some light on the condition of the machine. Dirt, discoloration, and acidity are reasons for a change of oil. Any excessive abnormal condition should be investigated for its root causes.

When operated under design temperature and load conditions, sleeve bearings are a very reliable part of the machine. However, severe vibrations, lack of sufficient flow of oil, deficient cooling, and external pollutants such as foreign materials and shaft currents can result in bearing failure.

Foreign materials tend to become embedded in the babbitt. Shaft currents tend to show up like minuscule spots of melted material. When viewed with the aid of a magnifying glass, depending on the angle of the light, they may appear as tiny rounded pits. For a comprehensive description of shaft currents and their control, see item S15 in Chapter 4. Polishing the journals and resurfacing (re-babbitting) the bearings are common operations during major overhauls. An additional activity related to bearing inspection is to check the integrity of bearing insulation and grounding brushes.

Working with large friction bearings is a delicate task requiring adequate expertise. It is thus highly recommended that only individuals with the proper experience and knowledge be assigned the task of inspecting and refurbishing the bearings of a large machine.

EPRI is sponsoring projects conducive to the application of magnetic bearings in large electric rotating machinery. When (and if) this technology becomes widely implemented, things will never be the same in the areas of bearing operation and maintenance.

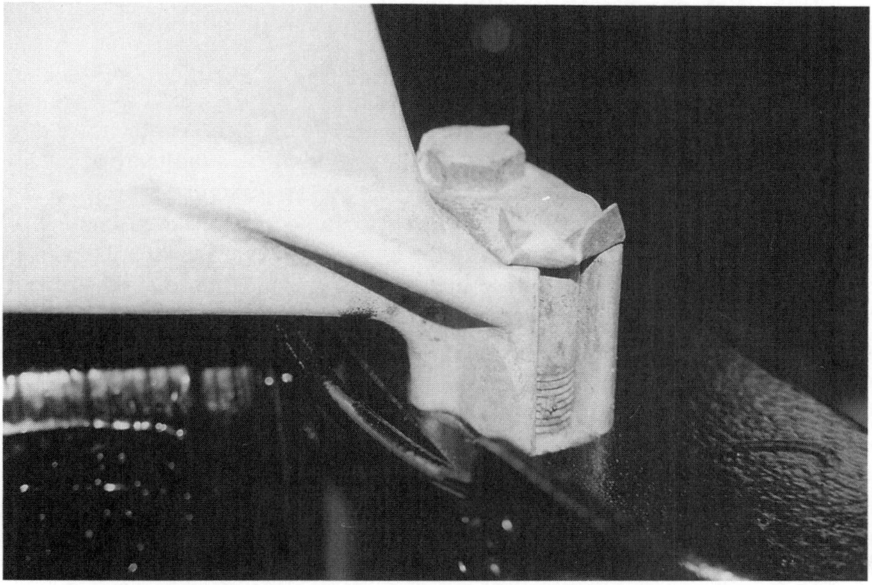

Fig. 5-10 [R06] Close view of one of the fan blades shown in Figure 5-8. The picture shows the root of the blade, the bolts attaching it to the rotor's frame, and the locking device that keeps the bolts from becoming loose during operation.

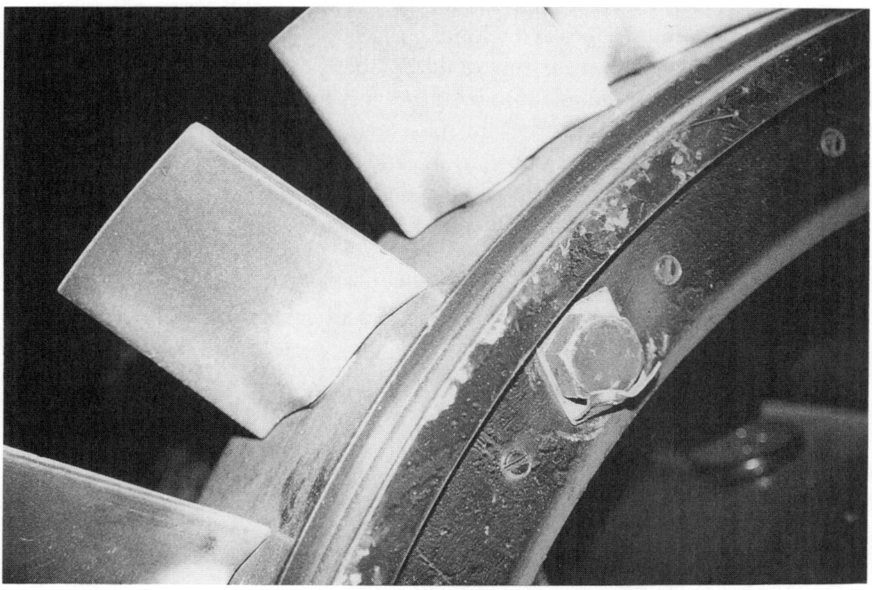

Fig. 5-11 [R06] Close view of the elaborate attachment of the fan blades in the rotor shown in Figure 5-6.

R08: Balance Weights/Bolts Condition

In large machines, balance weights, bolts, nuts, and any other rotor attachments are subjected to intensive centrifugal forces. For example, a 2-ounce nut will exert a centrifugal force of over 100 pounds when attached to the rotor of a typical 2-pole machine. In addition, the constant vibrations and thermal cycles tend to work these loose. A weight, nut, or bolt broken loose under these conditions often results in a major failure of critical components of the machine. For these reasons, these elements are secured with many types of locking gadgets or arrangements. It is important during the inspection of the unit to verify that all such nuts, bolts, and weights are secure and all locking devices are in order. This is particularly important when work has been performed on the rotating member during the overhaul.

R09: End-Wedges Condition (RR)

During unbalanced load or supply conditions (generator or motor mode of operation), as well as during system oscillations or other types of abnormal operation, alternating currents are established in the body of the rotor. These currents tend to flow along the rotor body (poles and teeth), along the wedges, and in the end bells (retaining rings).

When bridging high-resistance contact areas, these currents may give rise to very localized pitting of the metal. These high-resistance areas are mainly found in the contact surfaces between the wedges and the slot, between different wedges, and between wedges and retaining rings.

It is possible that severe pitting of the end bells can lead to catastrophic failure due to crack growth initiated from pitting. It is for this reason that wedges are kept from touching the end rings in most machines. Although normally pin-locked, wedges tend to migrate toward the end rings during the operation of the machine. If this situation is encountered during inspections with the rotor removed, the offending wedges should be driven back away from the rings (see Figs. 5-12 through 5-14).

After severe motoring or generation with the field off (induction mode operation), inspection of the rotor should always include inspection of the wedge-ring contact area, as well as between the wedges themselves and between wedges and the rotor body. It is important to note that any sign of burning should be carefully investigated.

Some manufacturers include a damper winding in 2- and 4-pole turbogenerators. These are designed to minimize the adverse effects of induced currents in the rotor. Nevertheless, these rotors should also be inspected for signs of induced currents.

Fig. 5-12 [R09] The narrow strips are rotor wedges of a cylindrical rotor. The picture shows how the wedges have migrated toward the end ring, with which they are in contact.

Fig. 5-13 [R09] Rotor wedges that have migrated.

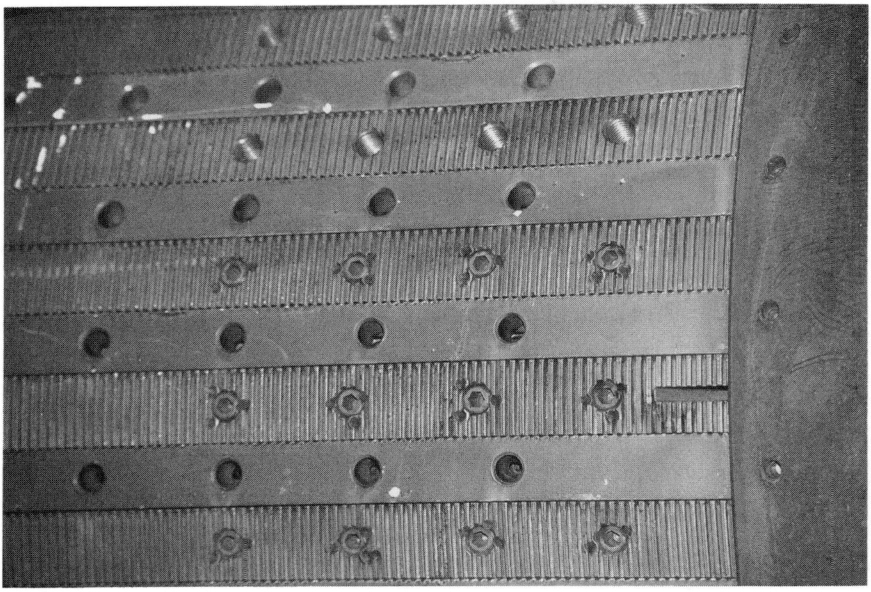

Fig. 5-14 [R09] A turbine generator's rotor having a single wedge per slot. In this machine, the wedges are designed to touch the retaining rings.

Other manufacturers design their rotor wedges to be equal to the full length between the end bells and to be in contact with them (see Fig. 5-14). These wedges, made of aluminum and short-circuited by the retaining rings, are designed to conduct the negative-sequence currents (when present), providing adequate damping of rotor oscillations.

R10: Other Wedges (RR)

Check all wedges for discoloration, which indicates the machine has been operated under abnormal conditions. Check for cracks and excessive looseness. If the clearances between wedge and slot appear to be excessive, contact the OEM for recommendations. The various designs and manufacturers differ in the number of clearances allowed between rotor wedge and slot.

Overheated wedges, in particular those made from aluminum alloys, might have distorted or lost their original mechanical properties. Consult the OEM for recommendations for each particular design.

In the case of aluminum and brass wedges, particular attention should be given to points of greatest mechanical stress concentration. Inspection of these can only be carried out with the wedge (and hence at least one end ring) removed. Obviously, this costly operation is only justified under very special circumstances (see Fig. 5-15).

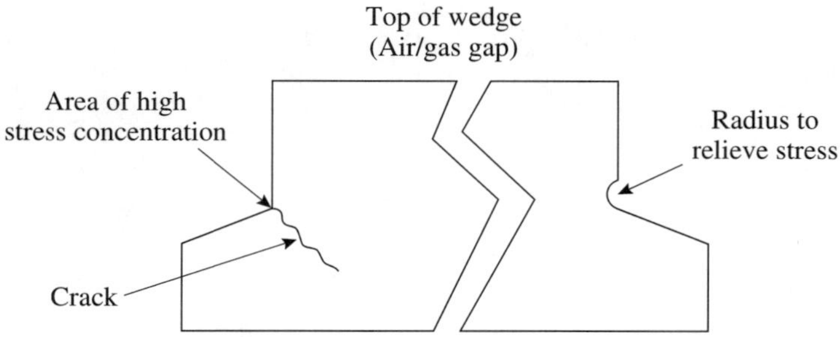

Fig. 5-15 [R10] Cross-section of aluminum rotor wedge showing stress area.

R11: End-Windings Condition (RR)

The rotor end-windings in a turbine generator are subject to intense centrifugal forces, as well as forces deriving from the cyclic changes in temperature. The radial displacement is contained by the end bells (retaining rings). However, a well-designed rotor allows for relatively free movement of the entire winding in the axial direction. It is important that, regardless of the axial movement of the whole end-winding, the distances between the individual coils underneath the end rings are maintained. The distances between individual coils are maintained by the introduction of preformed insulating blocks. Figures 5-16 and 5-17 show two views of the end-winding of a 2-pole cylindrical rotor.

However well designed, the end-windings of turbogenerators tend to deform. The degree depends on the machine's operating age, the type of loading (cyclic or continuous), any abnormal operating conditions encountered (short circuits, power swings, etc.), and the actual physical design of the coils. If the distortion is excessive (i.e., the individual turns are not aligned in each coil), then they become prone to develop shorted-turn faults. This type of fault is more likely to occur in DC fields in which the individual turns in the end-winding section are not fully insulated, but are protected only by a single layer of insulation between the turns. Excessive deformation can also result in cracks or breaks in the conductors, in particular in those coils having a sharp 90-degree angle in each corner.

Experience shows that one of the most onerous modes of operation to the integrity of the field winding in turbogenerators is prolonged low-speed rotation of the rotor ("on turning gear"). This operation is designed to eliminate the permanent bending of the shaft that would otherwise occur if the machine were kept still for long periods of time. However, the low-speed rotation results in continuous pounding of the windings due to their own weight, in particular the end-winding regions of the field winding. This continuous pounding is exacerbated if the clearances between the winding and the insulation are excessive. The result of

Fig. 5-16 [R11] Front view of the end-winding of a 2-pole cylindrical rotor having the retaining ring removed. In this picture, the top connection between each pole winding can be seen.

Fig. 5-17 [R11] Top view of the same end-winding shown in Figure 5-16. The blocking between coils has been removed. As seen in the picture, there are seven coils per pole in this example.

this continuous pounding is the creation of copper dust that may result in short-circuiting turns of the winding or even provide a path to ground in the slot region of the winding.

In certain designs and under certain operating conditions, the top turns of the field winding tend to crack and eventually break in the region just outside the slot. This type of failure can result in excessive damage to the rotor because of major ground faults developing as a result of the broken conductors.

Finally, the weakest point on the copper leads connecting the collector rings to the end-windings appears to be at the elbow of the lead, where it emerges from the shaft and rises to connect to the coils.

Inspection of the end-windings is not simple. They are quite inaccessible. They are confined between the retaining rings, shaft, and rotor body. For the most part, they are surrounded by insulation blocks. However difficult, a good and patient effort will most probably yield enough information on the condition of the winding to determine whether the machine requires additional inspections or is fit for operation. The most useful tools are the mirror and the boroscope. Check for alignment of the coils as well as turns within the individual coils. Check for loose blocking (Fig. 5-18) and cracked turns. In particular, try to concentrate on the top turns close to the slot. Also pay attention to the condition of the leads at their elbow. Look for excessive copper dust. If this is present, the inspector may rec-

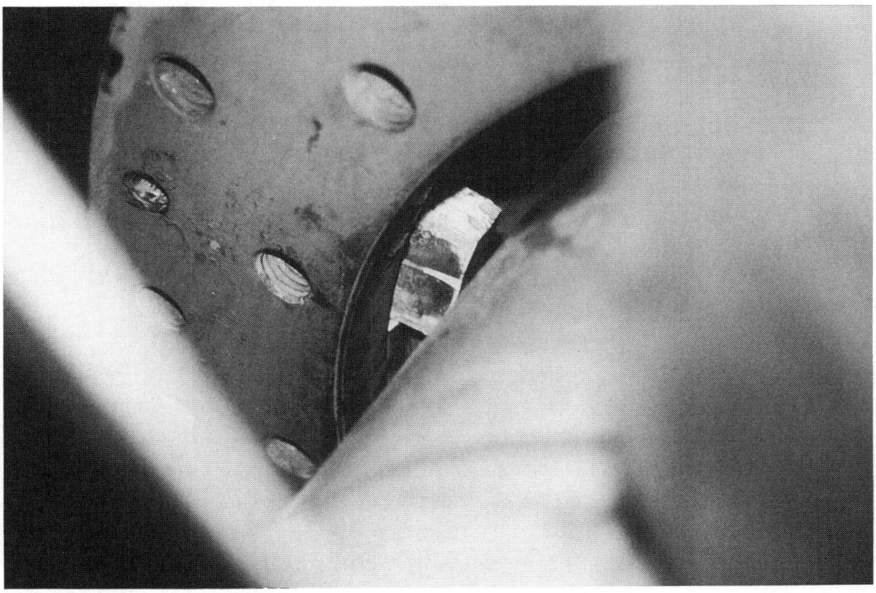

Fig. 5-18 [R11] Loose insulating blocking of the end-winding of an air-cooled gas-turbine generator. The blocking skidded from its position between the coils under the retaining ring, and it is seen sitting on the shaft.

ommend an electric test to check for the presence of shorted turns in the field. Chapter 8 sheds some light on available methods for detecting shorted turns. If deemed necessary, close inspection of the leads might require the removal of the lead wedges. This entails some additional disassembly.

When excessive copper dust is encountered, removal may be tried by blowing, or sometimes by washing with steam or liquids. Although the practice of using liquids or steam in turborotors exists, some experts are wary of applying it. The same, to a lesser degree, applies to dry blowing. Some experts believe these procedures result in carrying the dust back into the slot region, where it could end up causing more damage than before. As an alternative, vacuuming may not remove all of the copper dust; however, it will minimize the risk of contamination of other vital areas.

R12: Top Series Connections (SP)

The field poles of a salient-pole machine tend to move relative to each other during the operation of the machine. Although the resulting displacement is small, its continuous occurrence results in metal fatigue of the top series connections between poles. For this reason, manufacturers use flexible connections on the top series connections. Nevertheless, over many years of operation some of these connections may show indications of metal fatigue, namely cracks or breakdown of some of the copper laminations making up the connection. In many instances the connections are covered by insulation. If the integrity of these connections is suspected, the inspector should have the insulation removed from at least a few connections to allow for an unimpaired visual inspection.

R13: Bottom Series Connections (SP)

Although less susceptible to damage originating from metal fatigue, bottom series connections should also receive attention from the inspector, as they represent a weak point in the rotor winding.

R14 and R15: Field-Pole Keys in Dovetail and Inter-Pole Blocking (SP)

In many salient-pole rotors the poles are attached to the forging by dovetails and secured with a single or a double wedge-key system. They should be checked during major inspections for looseness. At the same time, the *V*-shaped inter-pole blocking should be inspected for looseness. These blocks support the pole windings, preventing distortion due to the centrifugal forces encountered in these machines (Fig. 5-19). Sometimes these are made of metal, and in many instances they are made of solid insulating material, which can crack over years of operation. Therefore, they should be inspected for cracks or broken parts.

Fig. 5-19 [R15] A *V*-block wedging two adjacent poles in a salient-pole generator. Insulation strips protect the coils from direct pressure by the *V*-block.

R16 and R17: Insulation between Turns (SP)

The salient-pole DC field windings come in essentially two distinctive designs: the *strip-on-edge* type and the *wire-wound* type.

The strip-on-edge type, encountered mainly in large hydrogenerators and relatively high peripheral speed machines, is made out of layers of copper strip joined at the corners or bent to form a multilayer coil around the pole (see Fig. 5-20). The ground insulation is located between the coil and the side, top, and bottom of the pole. In older machines, the ground insulation is made of mica, asbestos, and fish paper bonded with shellac or other organic materials. In recently manufactured machines, asbestos has been eliminated from the list of allowed insulation materials. The coil is insulated from the top and bottom of the pole with insulation materials having high mechanical compression qualities. In the past these were mainly asbestos boards; nowadays they are made of materials other than asbestos, with similar electrical and mechanical qualities.

The wire-wound pole type, used mainly in smaller and slower machines, is made out of preinsulated magnet wire wound in layers of multiple turns (see Fig. 5-21). The wire or turn insulation is made of aromatic paper, resin-bonded glass, mica-glass, and so forth. The insulation between layers and between layer and pole is of the kind described previously for the strip-on-edge type of winding.

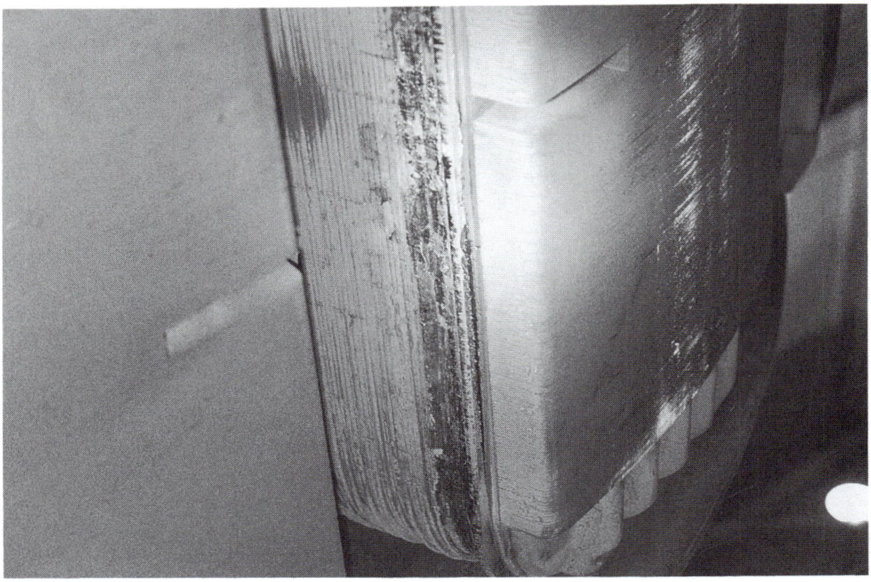

Fig. 5-20 [R16] A salient pole with a strip-on-edge winding.

Fig. 5-21 [R16] A wire-wound salient pole.

The main problems associated with salient-pole windings are due to the large centrifugal forces acting upon the windings. These forces tend to distort the conductors. In addition, the continuous vibrations and movement of the poles, associated with the radial clearances between pole and winding resulting from the centrifugal forces, tend to result in abrasion of the wire and/or turn insulation. Spring-mounted coils are designed to minimize these movements and abrasion by keeping a positive force acting at all times on the winding.

Visual inspection should search for coil deformations, for insulation dust indicating excessive movement between layers (not enough pressure), for broken or cracked collars (washers), and for broken or displaced springs (if present).

R18: Starting-Bars (Damper Winding) Condition (SP)

Salient-pole machines are frequently fitted with a starting winding. This can take the form of a squirrel cage winding occupying the entire circumference of the rotor or, as with most large machines, the starting winding is restricted to the pole regions. At the poles, the cage bars are imbedded in the face of the pole. These bars, which are short-circuited at the ends, function like an induction motor's squirrel cage during start-up operations, allowing the machine to start from zero speed and go up to near full speed, without the need of "pony" motors or variable frequency drives. In addition, the bars provide electrical damping to oscillations during the synchronous operation of the machine.

The starting (amortisseur) winding is designed in accordance with the chosen mode of starting. For example, some machines are designed to start at reduced voltage, while others start at full voltage. Operator errors, such as higher-than-designed voltage starting, starting too often, prolonged asynchronous operation, and other abnormal conditions, can result in overstressing of the starting winding.

Figures 5-22 and 5-23 show the consequential damage from a failed starting squirrel cage of a 60-MVA synchronous condenser. The short-circuiting ring on one side of the cage broke loose from the bars, completely destroying the entire end-winding on that side of the stator winding. This catastrophic failure requires changing the entire stator winding.

Visual inspection is very effective in this case to assess the condition of the damper bars. Discoloration and/or deformation attest to abnormal operation. The bars should be checked for cracks or breaks. In some cases, removal of the paint in the region of the junction between the bars and the short-rings allows for a more effective visual inspection. If hairline cracks are suspected, nondestructive-tests (NDT) with penetrating dye ink should be performed.

Fig. 5-22 [R18] The short-circuiting ring of the starting winding of a 60-MVA synchronous condenser. The short-circuiting ring has broken from the conductor bars and damaged the stator winding in the entire circumference.

Fig. 5-23 [R18] A close view of Figure 5-22, showing a pole-face with the starting bars sheared off the short-circuiting ring. The damage to the stator winding can also be seen at the top of the picture.

R19 and R20: Bull-Ring Segments and Bracing to Starting-Bars (SP)

Bull-ring segments, or short-circuiting rings, are the electrically active elements to which the starting-bars are braced at both ends of the machine (Fig. 5-24). As explained previously for the bars, the short-circuiting segments should be inspected for cracks and overheating. In particular, the junctions of the rings and the bars should be closely inspected. This is the region most prone to failure in the starting winding.

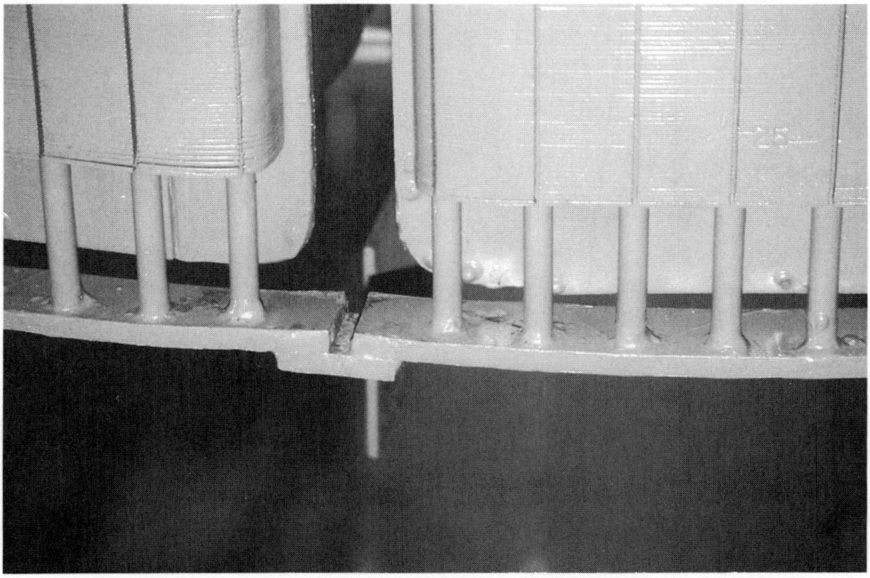

Fig. 5-24 [R20] Short-circuiting segments.

R21: Collector Rings Condition

The brush-rig and the collectors or slip-rings are probably where most of the wear and tear of a synchronous machine occurs (Fig. 5-25). Although simple in construction and appearance, the physics of the transfer of electric current at the brush-collector contact surface is rather complex. Under some conditions, the brushes and the collector would wear very little over many hours of operation, while under other environmental conditions, the same brushes and/or collector will wear overnight. The conditions that affect brush-collector performance are:

Fig. 5-25 [R21] The collector ring (slip-ring) of a large turbine generator.

- Level of humidity
- Contamination (gases, solids, or liquids, particularly carbon and/or metal dust)
- Current density
- Ambient temperature (too low or too high) and cooling of the collector and brushes
- Change of brush grade, wrong grade, or multiple grade
- Altitude (barometric pressure)
- Unevenness of collector surface
- Brush pressure
- Brush snugness inside brush-holder
- Insulation condition at bottom of collectors
- Poor shunt connections

In addition to the above direct elements affecting the operation of the brushes, there are other ways in which the performance of the brush-collector is influenced. For instance, a solid-state source of field current with poor commutation characteristics (i.e., severe harmonic content) results in accelerated wear of the brushes. Similar problems might arise from poor commutation in rotating exciters. Machine vibration also affects the performance of the brushes. Although vibration levels in electric machinery are kept within limits due to other considerations, it is convenient to remember that high vibration levels (more than 1.5 mils for 3600-rpm and 2.5 mils for 1800-rpm machines, approximately) will tend to have a detrimental effect on the operation of the brushes.

Another cause of bad brush-collector performance is axial misalignment of the brushes against the collector. This misalignment can arise from a combination of thermal expansion of the rotor, rotor axial movement off its magnetic center during operation, and wrong initial positioning of the brushes. To avoid this situation, proper axial positioning of the brushes against the collector rings should be visually corroborated when the machine is both cold (after start) and hot (after several hours of operation at full load).

A good or bad performance of the brushes depends on the condition of the contact film (called the *patina*). This film is made of carbon and copper and/or metal-oxide particles, elements present in the surrounding atmosphere, and water vapors. The film is primarily dependent on the collector metal being vaporized by the flow of electricity, then oxidizing and combining with water. This conducting film lubricates the collector, permitting the brushes to slide with minimum of wear. However, the integrity of the film is very fragile. Thus, the conditions required for a good film should be maintained as much as possible. Very low temperature, contaminants, and other parameters detailed previously are likely to affect the condition of the film. Attention should be paid to substantial changes of any one of these conditions, and their possible effect on the performance of the film should be evaluated.

Given the volatility of the brushes' performance, in particular for machines in which the brushes and collectors are exposed to the environment, a cursory inspection should be made every day the machine is running. More detailed inspections should be carried out in weekly or biweekly periods, or in particular when the machine is standing still or in turning gear.

Close attention should be paid to the condition of the brush-rig and collectors during a major overhaul inspection. At this time, evaluation of the collector's condition will indicate if "turning" or polishing of the collector's surfaces is required, or if any action should be taken to improve the insulation of the collectors, or to replace the collectors due to insufficient thickness or insufficient groove depth (for collectors with cooling-grooves), or if rotation of the polarities is indicated. It is normal for brushes to wear faster on the positive collector ring and slower on the negative ring. This property is used to defer the replacement of the slip-rings by changing polarities periodically (for instance, during major overhauls). This polarity dependence is more accentuated in condensers, where the collectors are also located (in many cases) within the hydrogen medium. Therefore, changing polarities is very common in these types of machines. With some designs in large turbogenerators, the expense of changing around polarities may not be justified by the resulting savings in the collectors.

When polishing the collectors, a minimum thickness of metal should be maintained to avoid disintegration by the centrifugal forces encountered during operation. The minimum thickness depends on such parameters as type of metal, diameter of the ring, speed of the machine, and calculated surface temperature. This author is not aware of a published formula applying to every type of machine design. The customary and proper action is to obtain data from the machine manufacturer regarding minimum recommended thickness, as well as minimum recommended depth of grooves (if called for in the design). When polishing or "turning" the rings, it is important that the finished surfaces do not exceed the recommended runout limit of 1 to 2 mils for 3600-rpm machines. Although higher runouts may be acceptable for lower-speed machines, it is recommended to remain within the 2 mils limit.

During extended periods in which the machine is inoperative, the brushes should be removed from the slip-rings and the slip-rings should be protected from humidity or other contamination. The same protection is necessary when the rotor is removed from the machine during large overhauls. In any case, the condition of the collector surfaces should be ascertained before returning the machine to operation.

During overhauls the condition of the brush pressure springs should be evaluated. Discoloration of the springs can be a sign of overheating. Overheating deteriorates the performance of the spring. Normal brush pressures are in the range of 1.75 to 2.25 psi of brush cross-sectional area. These, like any other values given in this book, are for reference purposes only. Actual values should be obtained from the brush manufacturer based on the grade of the brushes, the speed of the machine, current densities, and so on. For machines in which overhauls occur at long intervals (5, 6, 7, or more years), the cost of replacing the brush springs could be more than compensated by the consequent savings in brush and collector materiel. The constant-pressure type of spring is preferable.

The dependence on so many factors often makes the reasons for bad brush performance appear extremely elusive. The combinations are too many to be described here. However, over the years, manufacturers of brushes have developed graphs and procedures as aids to solve problems in a step-by-step approach.

R22: Collector Insulation Condition

The collectors (slip-rings) are mounted on a layer of insulation material. This insulation is subject to contamination by oil, water, carbon dust, copper, iron and other metal particles, and other chemicals. If those contaminants are not periodically removed, an insulation breakdown may occur. This can show up as a grounded field, or, worse still, may result in a severe short circuit of the DC field current. During major inspections, it is important to spend some time assessing the integrity of this insulation.

R23: Brush-Spring Pressure and General Condition

As stated in item R21, normal brush pressures are in the range of 1.75 to 2.25 psi of brush cross-sectional area. However, actual required values should be obtained from the brush manufacturer or the machine's OEM. The required pressure depends on the brush grade, humidity, current density, peripheral speed, and so forth. Once the desired pressure is known, it is important to make certain the springs are providing it on a continuous basis. Constant-pressure springs are preferable. Discoloration of the springs may indicate a weakened brush. Refer to item R21 for a more in-depth commentary.

R24: Brush-Rig Condition

The brush-rig is insulated from ground. During major inspections this insulation should be examined. The same type of contamination to the insulation of the slip-rings as discussed in item R22 applies to the insulation between the brush-rig and frame. During overhauls it is often necessary to remove the brush-rig in order to disassemble the machine. Thus, integrity of the brush-rig and its insulation is easy to evaluate under those conditions.

R25: Shaft-Voltage Discharge-Brush Condition

In item S15, a detailed account of the dangers to the integrity of the bearings due to shaft voltages is given. It is the practice in large electric machines to install discharge brushes to avoid the buildup of excessive shaft voltages. The condition of these devices (normally brushes of different types—see item S15) should be evaluated often.

R26: Inner/Outer Hydrogen Seals (RR)

In hydrogen-cooled turbogenerators with brush-fed excitation, the leads between the DC field winding and the collector rings are routed through the interior of the shaft; i.e., portions of the shaft are hollow. Were specific precautions not taken, the same space would allow the pressurized hydrogen in the machine to escape. Escaped hydrogen has been known to mix with oxygen and cause explosions in the brush-rig compartment due to arcing of the brushes or other igniting source. This directly poses a potential safety hazard to personnel.

To eliminate the possibility of hydrogen leaks, so-called *hydrogen seals* are installed. Often two sets are installed, one in the inner part of the rotor, where the leads leave the shaft and connect to the field winding, and another outside the casing, beneath the collector rings, where the connection between those and the leads is located. There are two seals in each location, one per lead. However, in some machines there is only one set of hydrogen seals, normally in the inner side.

It is practically impossible to assess the condition of these seals visually unless a significant dismantling of rotor parts is carried out. Even then, the final and definitive test is a pressure test carried out by fabricating a specially designed cylindrical can, which fits snugly over the shaft extension and seals the area of the slip-rings. This can is then pressurized exceeding the maximum operating pressure, and the pressure is monitored for a number of hours. If it is maintained, then the integrity of the hydrogen seals is demonstrated.

R27: Circumferential Pole Slots Condition (RR)

In certain round-rotor designs, a number of circumferential slots are machined in the poles of the forging a few inches apart from each other (see Fig. 5-26). The purpose of these slots is to introduce a measure of flexing freedom to the rotor by stiffness equalization in the pole axis, to minimize vibration. The tips of the slots, where they reach the outside diameter (OD) of the rotor—in the face

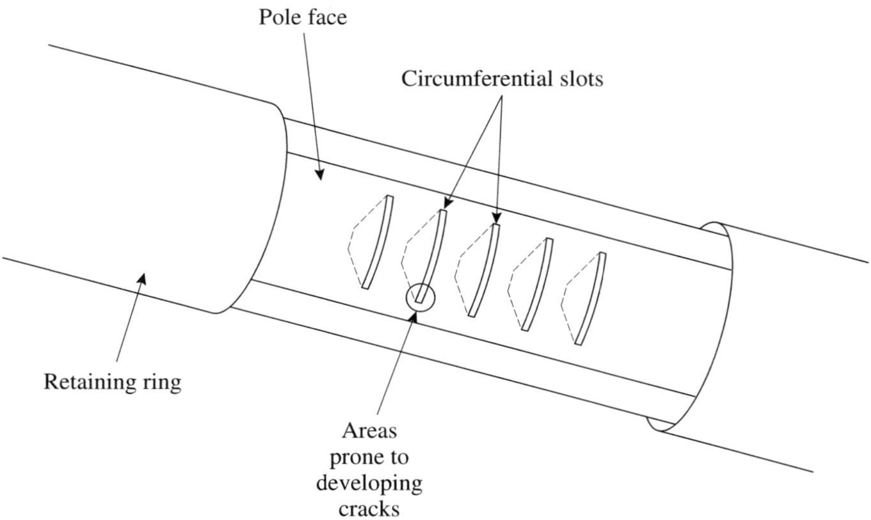

Fig. 5-26 [R27] Schematic representation of the circumferential pole slots, indicating areas prone to developing cracks.

of the pole—are areas prone to develop stress cracks under certain abnormal operating conditions. For instance, excessive asynchronous operation (without the DC field current) might result in currents flowing through the body of the rotor for extended periods. In this case, the tips of the slots become areas carrying high current densities. The highly localized heat generated by the high-density current flowing while these abnormal conditions exist may result in localized changes to the metallurgic properties of the steel. Consequently, the more brittle metal may develop cracks due to metal fatigue, as the rotor flexes at twice-supply frequency.

To avoid the possibility of catastrophic crack growth, the inspection team should be aware of any signs of overheating in those areas (normally appearing as discoloration of the metal). If encountered, a nondestructive test or examination (NDT or NDE) should be performed. Any crack found should be cause for further investigation and the initiation of adequate remedial measures.

NOTE

[1]*Turning gear* is the term given to the motor and associated gearbox used to turn the rotor at low speed, when the machine is not in operation. Turning the rotor at low speed keeps it from creating a bow due to its own weight. In particular, it is important to turn the rotor immediately after bringing the machine off-line, when the rotor is still hot; a thermal bow can be excessive if the machine is allowed to rest.

REFERENCES

[1] EPRI Special Report EL/EM-5117-SR, "Guidelines for Evaluation of Generator Retaining Rings," April 1987.

6

Description of Excitation Items

Chapter 6 describes each item on Form 7, the Excitation Inspection form. The items in Form 7 attempt to cover, however generally, the various types of excitation systems that are typically employed in large synchronous machines, whether self-excited, stand-alone solid-state, shaft-driven generator, stand-alone generator, or some other system.

E01: Cleanliness

Items R21 to R23 in Chapter 5 stressed the sensitivity of the operation of the collector brushes to the brushes–spring–collector condition, as well as to the suitability of the environment, which includes surrounding atmosphere and dirt or other contaminants. Therefore, excessive presence of these contaminants is a plausible indication of less-than-effective operation. If excessive carbon dust or copper dust is present, it could be an indication of too much wear on the brushes and collector. If left untreated, a large amount of dust can restrict the movement of the brushes in their brush-holders, resulting in a quickly deteriorating situation. Excessive copper and/or carbon dust may also result in a ground fault on the collectors on the brush-rig.

In short, the accumulated contaminants may provide a clue to the operation of the machine; however, contaminants should be removed as often as necessary to allow for effective brush operation.

Cleanliness obviously applies to other equipment associated with the excitation system. For instance, when free-standing solid-state controlled rectifiers provide the field current, they should also be opened and, as a minimum, given a cursory inspection. The soundness of the main connections should be verified. Filters, if present, should be cleaned or replaced, and the condition of the cooling fans, if installed, should be evaluated.

E02: Shaft-Mounted Diodes Condition

Many smaller synchronous motors and generators, and an increasing number of larger machines, are being supplied with shaft-mounted excitation systems. In this arrangement, a shaft-mounted auxiliary winding is energized by a controlled flux created by a stationary winding. The rotating winding produces alternating currents, which are then rectified by a shaft-mounted solid-state rectifier bridge. The rectified currents then flow into the DC field winding through a set of leads. The bridge consists of diodes mounted around the perimeter of the rotor in a diode-supporting rig.

Experience shows these diodes are very reliable. Testing the diodes may not be required in most cases. However, *it is important to disconnect both leads from the diode bridge if any potential test is to be carried out on the field winding.* Otherwise, the voltages generated by equipment such as hi-pot and/or megger devices will destroy the solid-state junction of the diodes. The diodes' heat-sinks should be cleaned if necessary to improve their cooling process.

E03: Diode Connections and Support Hardware

Although the diodes themselves are very reliable and may not need to be electrically tested, the connections between the diodes and leads, as well as the soundness of the supporting rig and its attachment to the rotor, are subjected to continuous vibration and mechanical centrifugal forces. Therefore, their soundness should be checked during a major inspection while the rotor is out of the bore.

Diodes are commonly protected by the diode fuses. These fuses normally have a pin that is released upon operation of the fuse. If such fuses are present, the inspector should ascertain that they are all operational.

E04 to E07: Commutator, Brushes, Springs, and Brush-Rig Condition

The commutator of the DC machine (in a rotating excitation system) should be inspected with the same scrutiny as the brush/collectors in the main machine. Most of the topics discussed for collector rings in items R21, R22, and R23 of Chapter 5 also apply to the commutator. In addition, special care should be taken to remove excessive accumulation of brush carbon and other dust in undercut mica grooves. This accumulation is intensified by the existence of oil contamination. Signs of arcing across mica, between adjacent segments, or between commutator and shaft indicate the need for cleaning. Hot spots should be noted and their causes investigated. Loosened and/or shifted segments should be noted and, if necessary, corrective measures should be taken. Unequal spacing of the brushes around the commutator may lead to defective commutation. If this situation is suspected, the position of the brushes can be measured by tracing their contour on a strip of thin paper placed between brushes and commutator. Subsequently, the paper is retrieved and the distances are measured. If necessary, corrections can be made by shifting the brush-holders on the brush-rig.

All the topics discussed in items R21, R22, and R23 of Chapter 5 regarding commutation film, brushes, springs, environment, and brush-rig support apply here entirely. Figure 6-1 shows a DC commutator with the brushes removed.

Fig. 6-1 [E04] DC generator commutator shown with the brushes removed from the brush-holders.

E08: DC Generator Stator Condition

If a rotating exciter is used, whether shaft-mounted or independently driven, it should be subjected to the same close inspection as that of the rest of the generator. It is a bad situation when an undermaintained motor–exciter system fails, paralyzing a very large generator in good operating condition. In general, stators and rotors of rotating exciters should undergo inspection procedures similar to those that apply to the main generator. Electric tests are also similar to those performed on the main generator. All major items for inspection described previously for the generator's stator apply to some degree to the stator of the exciter.

E09: DC Generator Armature Condition

Apart from general appearance, a visual inspection should assess the soundness of the connections to the commutator segments. The insulation should be analyzed for an indication of overheating, excessive dryness, movement, and so on. Figure 6-2 shows the armature of a shaft-mounted exciter of a large generator.

Fig. 6-2 [E09] Armature of shaft-mounted exciter.

E10 and E11: Exciter-Drive Motor Cleanliness and Stator Condition

Occasionally, large turbogenerators have the excitation drawn from independently driven DC generators. The drivers of these machines are induction motors with ratings up to several thousand horsepower. These high-rating motors are not available as "off-the-shelf" purchased equipment. Therefore, a catastrophic failure of such a motor may paralyze a major generating unit for long periods of time. To avoid these costly outages, the exciter's driver should receive the same attention as the main unit, although this is seldom the case. Inspection of the driver's stator is made along the lines of the items discussed in Chapter 4 for the main unit.

E12: Exciter-Motor Rotor

Practically all exciter electric drivers are of the squirrel-cage construction. This is a very rugged and reliable rotor. The major point for inspection is the integrity of the joints between the rotor bars and the short-circuiting rings. Any signs of discoloration due to high temperature should be investigated.

E13: Field Discharge Resistor Condition

In many synchronous machines equipped with breakers in the main field leads, the field is discharged across an external resistor when the main field breaker is opened. The resistor eliminates the high voltages induced in any large inductance when the current flowing through it is abruptly interrupted. From time to time, the integrity of this resistor should be confirmed by inspection. A bad resistor can be the cause of expensive field winding repairs.

7

Generator Auxiliaries

A major inspection of a generator would not be complete without careful inspection of the ancillary equipment (auxiliaries). In fact, failure of any auxiliary system can potentially result in a serious and long disruption of the operation of the unit and perhaps catastrophic failure of the machine.

Although the excitation system can be considered an auxiliary to the main machine, it has been covered above as a separate item (Chapter 6).

Auxiliary systems tend to vary to a great extent depending on the type of machine, manufacturer, rating, type of prime mover, and so on. Therefore, only components that are common to most systems are discussed in this chapter. The inspector of a particular system should refer to manufacturer instruction manuals, when available, for detailed specific information.

7.1 LUBRICATION SYSTEM

In many hydrogen-cooled generating units, the lubrication system is designed to provide a sufficient and reliable supply of oil to both generator and turbine, as well as to provide seal oil for the generator shaft's hydrogen seals. The typical lubrication system comprises a main oil reservoir, pumps, oil coolers, vapor extraction equipment (for the oil of the bearings' circuit), valves, and instrumentation (see Fig. 7-1).

Normally, the reservoir is located at an elevation below the unit deck so that oil can drain from the bearings by gravity. (In vertical hydrogenerators, the large

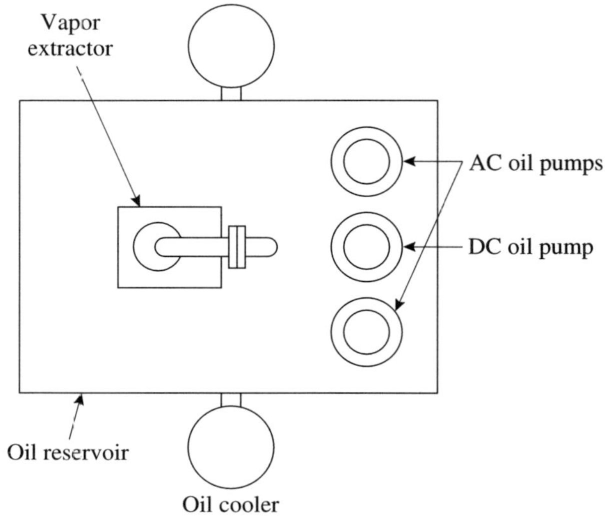

Fig. 7-1 Typical arrangement of a lubrication system.

thrust bearing is immersed in an oil tank, which may include a cooling heat exchanger.) The tank is sized to contain all the lubricant required for the reliable operation of the unit. Commonly, a motor-driven vapor extractor creates a negative pressure inside the oil reservoir. This causes an inward flow of air through the oil deflectors in the bearing house, eliminating oil leakage through the oil deflectors.

The oil is normally pumped by one or more AC motors. However, large machines have an emergency DC motor pump that automatically goes into operation when it senses a drop in oil pressure due to failure of the AC pump(s).

The oil is cooled by water–oil heat exchangers, generally mounted close or on the main oil reservoir. In most cases, AC pumps and coolers are designed with redundancy; i.e., the unit can operate with one pump and cooler while the other undergoes repairs or reconditioning.

During an overhaul, all components of the lubrication system should be inspected (including instrumentation, alarms, etc.). In actuality, most manufacturers recommend extensive inspection of the lubrication system at shorter intervals; e.g., every two years.

The inspection may include, but is not limited to, the following items:

- Tank cleanliness and overall condition
- Oil coolers: leaks, corrosion, cleanliness
- Oil pumps
- Vapor extractor
- Piping connections

- Pressure switches and gauges
- Fuses, contactors, and cables to pumps and other equipment
- Valves
- Gaskets

7.2 HYDROGEN SEAL OIL SYSTEM

Hydrogen-cooled machines exhibit oil systems designed to contain the high-pressure hydrogen gas within the machine. For all static joints, appropriate gaskets will do the job. However, the clearances around the shaft and the casing are sealed by means of a seal oil system located in each bearing.

The seal oil can be part of the main oil system, or contained in a reservoir of its own. Typically, the seal oil flow circuit comprises the air-side oil feed, cooler, and pump, and the hydrogen-side oil feed, cooler, and pump. The seal oil system also includes the hydrogen drain regulator, valves, and pressure sensors.

Inspection of the H_2 seal oil system includes the following items, at a minimum:

- Hydrogen seal insulation
- Ring segments
- Ring-shaft clearances
- Hydrogen seal casing
- Hydrogen seal regulator (several filters, gauges, flowmeters, etc.)

7.3 STATOR WATER-COOLING SYSTEM

Stators cooled by direct flow of water in the conductors require careful inspection of the water-cooling system inside the machine. This topic was covered to some extent in the items relating to the stator winding in Chapter 4. In addition, the inspector should carefully examine all those external components of the stator water-cooling system. This inspection includes the following items, among others:

- Filter and filter-pressure alarms
- Piping condition
- Water tank condition
- Pumps
- Demineralizers
- Conductivity meters
- Condition and cleanliness of heat exchangers

- Pressure regulators, valves, instrumentation
- Hydrogen leak sensors

7.4 HYDROGEN SYSTEM

In hydrogen-cooled machines, inspection of the hydrogen system represents an important step in maintaining reliable and safe operation of the unit. Inspection of the hydrogen system should include examination of the CO_2 system. The inspection should include, at a minimum, the following items:

- Integrity of all valves, pipes, flowmeters, and instrumentation
- Water detectors
- Gas dryer (blower and heater)
- Purity meters

8

Standard Electrical and Mechanical Tests

A substantial number of different electrical and mechanical tests are available for use during the inspection of a large synchronous machine. The choice of the tests to be performed depends on the scope of the inspection and the idiosyncrasy of the inspector(s). Some tests are never used by some inspectors but may be widely used by others. Some might be suited for the factory, some for the field. Tests that for all practical purposes can only be performed in a factory setting were not included herein. What follows is a list of those tests more widely used. The list is intended to serve as a catalogue rather than to provide information about how to perform the tests. For more details on the philosophy of the tests and how to perform them, the reader is referred to the pertinent standards, some of them quoted in the Reference section at the end of this chapter.

Tests, particularly high-voltage tests, should be carried out only by experienced and qualified personnel closely following stringent safety rules [1].

It is important to plan, before a major overhaul, which tests should be performed immediately before shutdown and which immediately after shutdown, before and after disassembly of the machine. For example, on-line PD readings and rotor-flux waveform readings may be taken prior to shutdown. DC resistance, megger, PI, and HI-POT, may be taken after shutdown. The objective of these tests should be to enter the visual inspection with enough test data to enable the inspector to focus on suspected problem areas, and to allow enough lead time for repairs, if needed.

8.1 ELECTRIC TESTS

8.1.1 Winding Resistance (DC) [2, 3]

During this test, the measurement of the ohmic value between the terminals of a winding is carried out. Given the relatively low series DC resistance of windings of large machines, the measurement's accuracy, to be significant, has to be to a minimum of two decimal places.

The purpose of the test is to detect shorted turns, bad connections, wrong connections, and open circuits. Acceptable test results consist of the three resistance values (one per phase) to be balanced within a 0.5% error from the average. The test is very sensitive to differentials of temperature between sections of the winding. The machine should be at room temperature when the test is performed. As with any other electrical test, the results should be compared with original factory data when available. This test can be performed on stator and rotor windings.

8.1.2 Insulation Resistance (DC) [4]

The purpose of this test is the measurement of the ohmic value between the conductors in different phases and between conductors and the iron core (normally grounded). Ordinarily, the measurements will be in the mega-ohm region after the winding is subjected to a DC voltage of a value of a fraction of the machine's rated voltage, for one minute. Every test in which significant voltage is applied will stress the insulation; therefore, the voltage test value should be chosen according to the condition of the insulation in the machine and the machine's voltage rating. Although the readings obtained will be slightly voltage-dependent, this voltage dependency is almost nil for machines with dry insulation in good condition.

The test is carried out with so-called megger devices. Resistance bridges are also employed. The use of a steady DC voltage supply, a voltmeter, and a low-current ammeter also allows the taking of resistance readings.

The readings are sensitive to factors like humidity, surface contamination of the coils, and temperature. Readings are corrected to a base temperature of 40°C with the following formula [5]:

$$R_{40^\circ C} = K * R_{measured\ in^\circ C}$$

where K is a temperature-dependent coefficient that can be obtained from Figure 1, page 9, of the ANSI/IEEE Std 43-1974 [5]. The following equation can be used to some degree of accuracy in lieu of the aforementioned figure:

$$K = 0.0635 * \exp\left[0.06895 * T_{measured\ in\ ^\circ C}\right]$$

The recommended minimum value of insulation resistance is a subjective issue.

However, most users of large synchronous machines in this country adopt the values recommended by the ANSI/IEEE Std 43-1974, section 9.2:

$$R_{\text{minimum}}\left[\text{in mega-ohm}\right] = kV + 1$$

where kV is the line rated voltage of the machine. The measured value to be compared with this value is the one-minute reading corrected for 40°C. When the three phases are measured one at a time, with the other two grounded, the measured value should be divided by 2. However, if guard circuits are used on the other phases, the measured value of each phase should be divided by 3.

Insulation Resistance (DC) tests are performed on both stator and rotor windings. Additional information on how to perform the test can be found in the pertinent publications [5–7].

8.1.3 Polarization Index (PI) [5]

The measured value of the resistance of the insulation is time-dependent as well as a function of its dryness (Fig. 8-1). In Figure 8-1, the horizontal axis indicates the length of time—in hours—the insulation was subjected to a drying process. In most cases, the rate of change becomes small after about 15 minutes. The amount of change in the measured resistance value during the first few min-

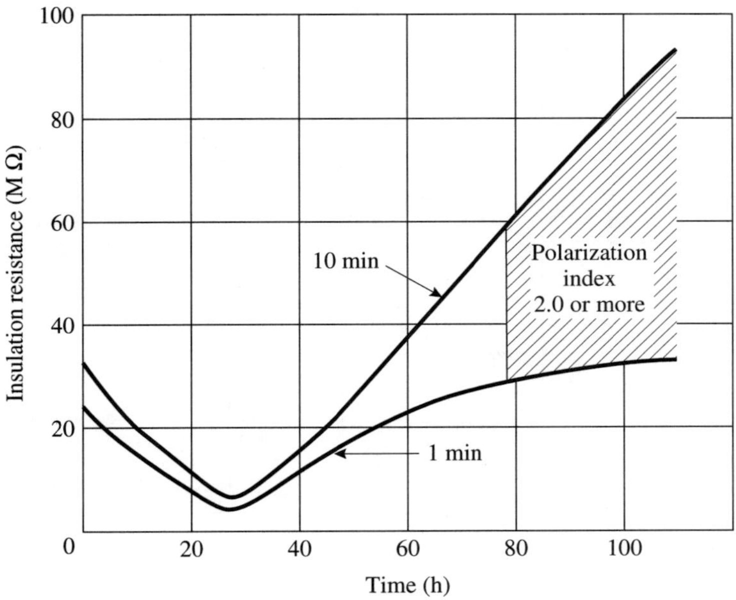

Fig. 8-1 Insulation resistance as a function of time and dryness. (Copyright © 1987, Electric Power Research Institute. EPRI EL-5036. *Power Plant Electrical Reference Series,* Volumes 1–16. Reprinted with Permission.)

utes depends on the insulation condition and external agents such as contamination and water. This behavior led to the introduction of the so-called *polarization index (PI),* which is the ratio between the resistance reading at 10 minutes and the reading at 1 minute.

The PI index is utilized to evaluate how clean and dry a winding is. The PI is also dependent on the winding components (Fig. 8-2). Class B windings tend to show higher PI values than windings made of Class A insulation. It is also dependent on the existence of a semiconducting layer. The recommended minimum PI values are [5]:

Class A insulation: 1.5
Class B insulation: 2.0
Class F insulation: 2.0

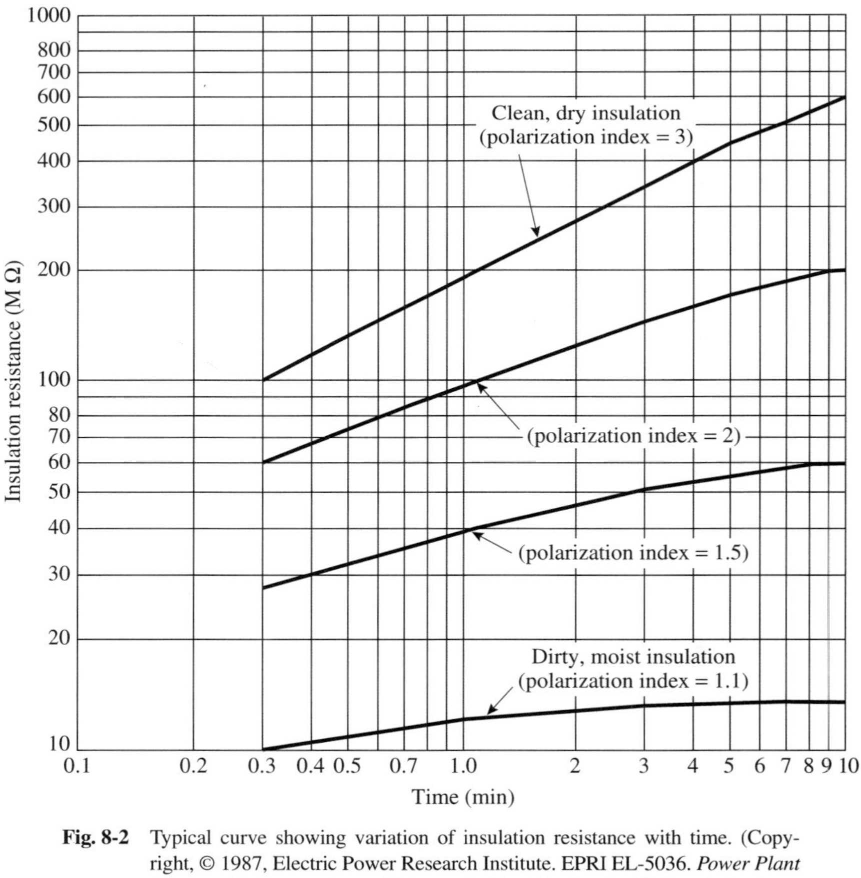

Fig. 8-2 Typical curve showing variation of insulation resistance with time. (Copyright, © 1987, Electric Power Research Institute. EPRI EL-5036. *Power Plant Electrical Reference Series,* Volumes 1–16. Reprinted with Permission.)

A PI reading is normally taken at reduced voltages as a go/no-go test before subjecting the machine to a high-voltage test (Hi-Pot). Performing the Hi-Pot on wet insulation may result in unnecessary failure of the insulation.

8.1.4 Dielectric (Over-Potential or Hi-Pot) Test [1, 8–11]

This is a test ascertaining if the winding is capable of sustaining voltage levels of the magnitude of the rated voltage without a breakdown of the insulation. The test consists of applying high voltage to the winding (the three phases together, or one at a time, with the other two grounded) for one minute. Depending on which variation of the test is selected, the applied voltage can be DC [10], AC at very low frequencies [9, 12–15], or at rated frequency. For a full description of the various tests, consult the referenced literature.

The recommended values for high-voltage tests range from two times the rated phase voltage plus 1000 V (rms) for factory acceptance tests, to slightly over the line voltage for some field tests. These values can be found in the referenced literature. In addition, the value actually chosen for the test voltage may also depend on the age of the machine's insulation and its general condition.

For the test of the DC field excitation windings on the rotor, the normally used values are between 1500 V to about 10 times rated field voltage [15].

8.1.5 Turn-to-Turn Insulation (Surge Comparison or Impulse) Test [16]

This test is designed to detect incipient breakdown of the inter-turn insulation in stator windings. Voltages used should generate potential between turns of the order of 10 times the normal voltage. The voltage imposed normally has a frequency of several kHz. The surges are produced by a choice of R, L, C components. Several tests have been devised, some specifically geared to test large synchronous machines [17]. Normally, these tests require the removal of the rotor.

The test is performed at different voltage levels. If a change in the waveform is observed, it means a discharge (breakdown of the insulation) has occurred. This test is almost always performed in individual new coils. Generally, when performed on a completed winding, its adequacy is accepted with certain reservations.

8.1.6 Shorted Turns in Excitation Field Windings [1, 18]

This test is designed to determine the existence of shorted turns in the DC excitation field winding. The test is entirely different when performed on salient-pole rotors than on cylindrical (solid) rotors. In salient-pole machines, a "pole drop" test is done. In this test, the resistance across each individual pole is measured by the V/I method; i.e., applying a voltage of approximately 100 to 120 V, 60 Hz, to the entire winding, and then measuring the voltage drop across each pole. A pole with lower voltage drop will indicate a shorted turn or a number of shorted turns. It is important to note asymmetries in the flux introduced by the pole with the shorts. This may affect the voltage readings in adjacent poles, giving slightly reduced readings. Oftentimes shorted turns are speed-dependent; i.e., they might disappear at standstill. To partially offset this phenomenon, it is recommended to repeat the pole drop test a few times with the rotor at several angles. The gravity forces exerted on the vertically located poles may activate some short circuits between turns that might not show up when in, or close to, the horizontal position.

In round rotors, the individual pole windings are for all practical purposes inaccessible, unless the retaining rings are removed. Therefore, detection of shorted turns in these rotors has always represented an elusive goal. Probably the most effective method for the detection of shorted turns in solid rotors is the flux probe method (see [19], Section 5.1.3.5). This device maps the flux of the machine as it rotates, indicating possible shorts as changes in the measured waveform (see Fig. 8-3).

Its main advantage is that it works with the rotor spinning, capturing the speed-dependent shorts. Its main disadvantages are the expertise required in analyzing the recorded waveforms and the fact that the machine has to be de-

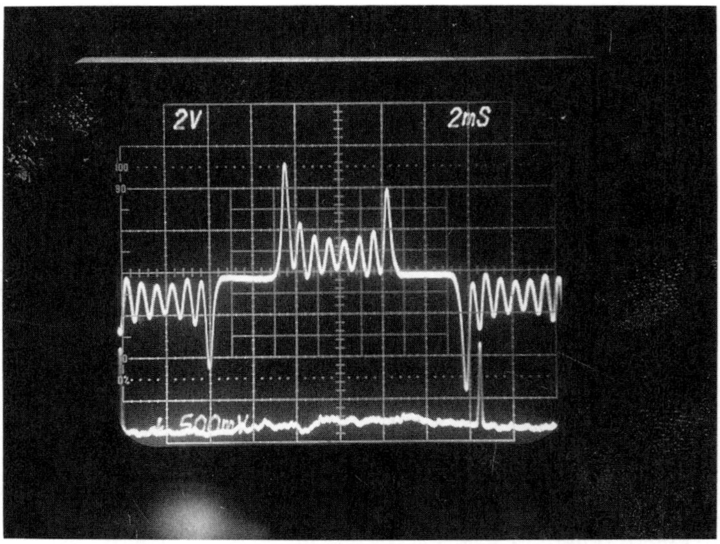

Fig. 8-3 Typical waveform recorded from a flux probe installed in a 2-pole turbine generator.

energized and degassed for the installation of the core-mounted type of probes, or partially disassembled for the wedge-mounted type. New commercially available units intended for on-line continuous operation include software that analyzes the waveform and alerts to a possible shorted-turn condition. Almost all other methods require the removal of the rotor.

The C-core test is one such method (see Figs. 8-4 and 8-5). To carry out the test, only a *C*-shaped wound core is required, together with a voltmeter, a wattmeter, and a single-phase power supply. Shorted turns are detected by sharp changes in the direction of wattmeter readings. In rotors with damper windings, or with the wedges short-circuited at the ends to form a damper winding, the wedges have to be disconnected at the ends. This operation requires removal of the retaining rings. Reference [19] further elaborates on this test.

One method that is currently being developed eliminates the need to remove the rotor [20]. It injects a low-voltage, high-frequency surge wave at each one of the slip-rings, and compares the same waveform observed at the other end. This method is currently being tried on spinning energized rotors; i.e., with the machine connected to the grid.

Impedance measurements while the machine is decelerating or accelerating can also be used to detect a speed-dependent shorted turn. Any sudden change in the readings may indicate a shorted turn being activated at that speed. A gradual change of impedance of more than 10% may also indicate a solid short (see Reference [19], Section 5.1.3.3 for threshold values and further details on this test method). Reference [19] includes several variations on tests for shorted turns.

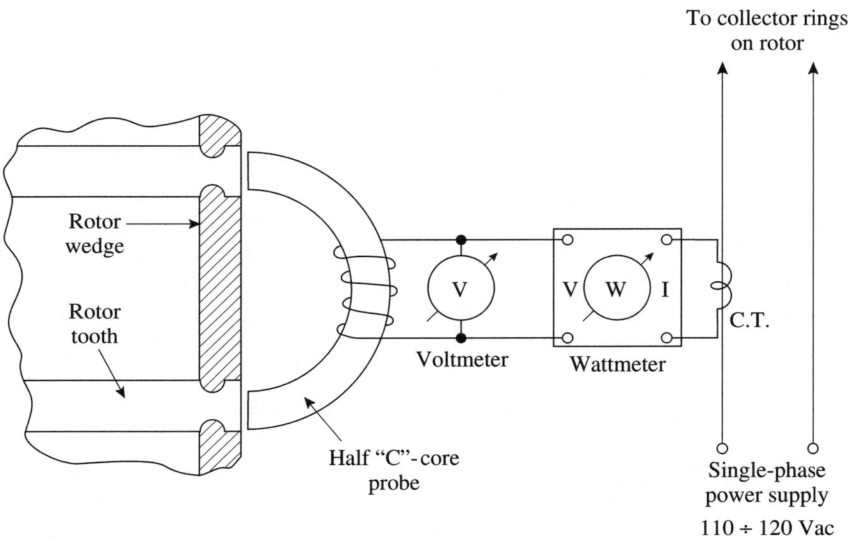

Fig. 8-4 Typical arrangement for the C-core test for the detection of shorted turns in the field winding of round rotors.

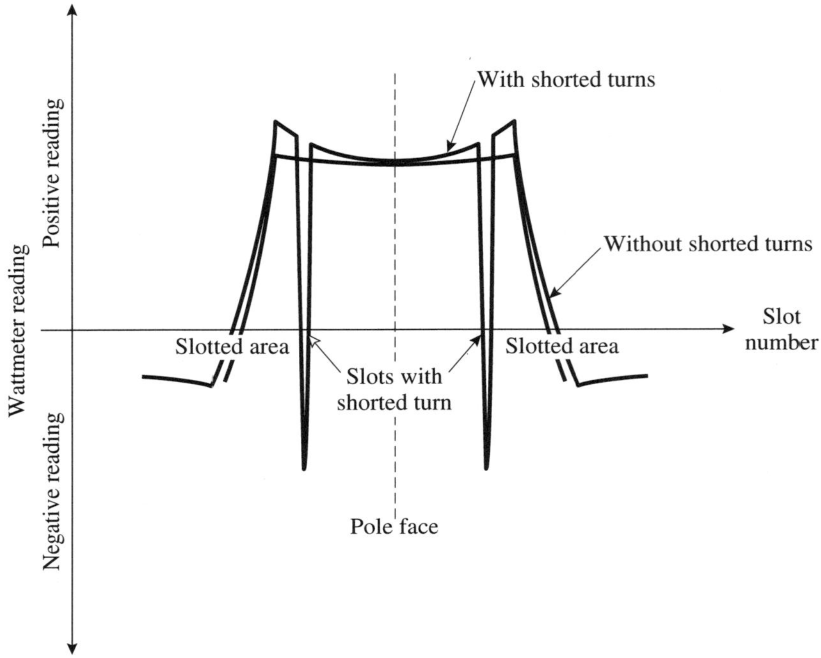

Fig. 8-5 Typical traces for the C-core test of Figure 8-4.

8.1.7 Open-Circuit Test (for Detection of Rotor Shorted Turns)

In all synchronous machines, open-circuit stator voltage versus field current characteristics can be measured. This curve, taken with the machine spinning at synchronous speed, is unique for each machine. Shorted turns in the rotor DC field winding may show up as a decrease of the measured voltage for a particular field current. However, due to the large number of turns in a typical rotor winding, the changes in open-circuit voltage due to a single shorted turn in the field winding may go unnoticed (too small for a positive identification).

8.1.8 Power Factor Test (for Stator Windings)

This test measures the dielectric loss per unit of volume of the insulation. It provides an additional measure of the insulation's condition. It is also dependent on the type of the dielectric material. One of the advantages of the power factor readings is that they are a dimensionless quantity. This quantity is generally expressed in percent.

An increase of power-factor in the readings over the life of the machine can be attributed to an increase in internal voids, and/or increased slot–coil contact resistance (deterioration of the semiconducting paint). A histogram of the insulation's power factor readings can prove to be a useful tool to assess the condition of the insulation.

Power factor tests are mostly confined to machines with rated voltages of 7000 V or higher. Insulation power factor readings are directly affected by the temperature of the winding. Thus, it is important that power factor readings be taken at similar temperatures. The power factor of the insulation is also a function of the voltage. Therefore, comparisons with previous readings should be made on tests taken at close voltages.

It has been described in the literature that insulation power factor readings are, to a certain extent, voids dependent; i.e., the power factor will increase with an increase in the number of voids present in the insulation. This phenomenon is the base of the power factor tip-up test (or simply "tip-up").

8.1.9 Tip-Up Test (for Stator Windings) [1, 21]

This test measures the density of voids in the insulation, as well as other ionizing loss, such as from corona or slot discharge. The test is based on the fact that ionization, both internal and external to the insulation, is voltage-dependent. By taking power factor readings at different voltages, a set of points are obtained, which are contained in an ascending curve. A fast change of power factor with voltage tends to indicate a coil with many voids (see Fig. 8-6).

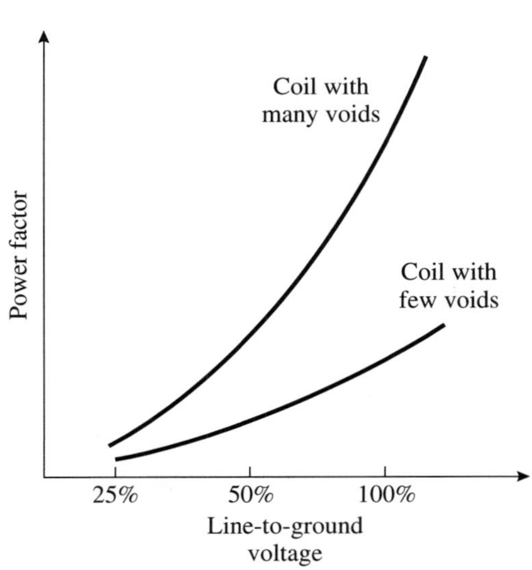

Fig. 8-6 Typical dependence of the insulation power factor on the number of voids in the insulation and applied voltage.

The test is done at 25% and 100% of rated phase voltage. The tip-up value is the percentage power factor at the higher voltage, minus the percentage power factor at the lower voltage.

Almost every manufacturer of high-voltage coils performs a tip-up test on individual coils to assess whether the manufacturing process left too many voids in the coil. When the test is applied to a whole phase, it will mask the individual coils. It will give a good evaluation to the winding as a group; however, any bad coil that deviates greatly from the rest will not be discerned by this test. This is probably one reason why many users do not use the tip-up test on connected windings. It is well known that the position of a coil with respect to the line coil affects the power factor reading (line coils are further prone to partial discharge due to the higher voltages to which they are subjected over the life of the insulation).

8.1.10 Dielectric Absorption [1]

This is a test designed to measure the aging of the bond in the cell insulation. When applying DC voltage to insulation material, a time-dependent flow of current is established (see Fig. 8-7). The current has a constant component, called the *conduction current* or *leakage current,* and a transient component, called the *absorption current.* The absorption current is a function of the polarization of the molecules in the bonding material. The older the bonding material, the more polarized it becomes and the more absorption current flows.

The dielectric absorption test is mainly a comparison of the winding condition at different times and between similar windings. The absorption current is also temperature-dependent, and this should be taken into consideration when performing the test and interpreting the results.

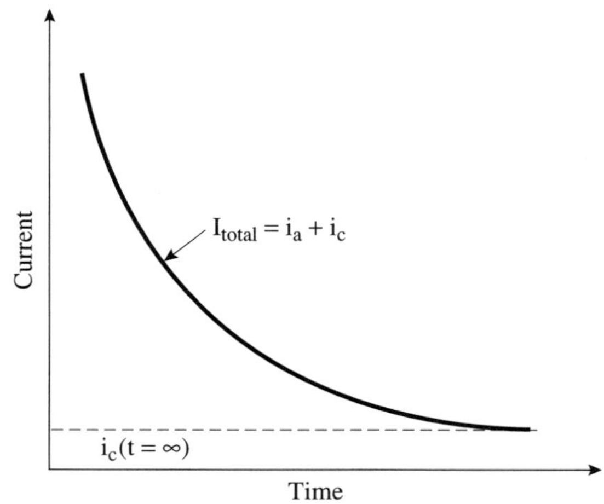

Fig. 8-7 Absorption current as a function of time (i_a = absorption current; i_c = conduction current).

Absorption current is also dependent on the number of voids in the insulation. The dependence is inverse; i.e., an increase in the number of voids in the insulation tends to reduce the magnitude of the absorption current. The contradictory effects of voids density and aging of the bonding material renders this test difficult to interpret. It is best when used in conjunction with other dielectric tests, such as *corona* tests and *tip-up* tests.

8.1.11 Partial Discharge (PD) Test (for Stator Windings)

The PD test measures the partial discharge activity within the insulation, indicating the number of voids in the insulation. In addition to tip-up, absorption, and other test methods, partial discharge can be measured directly by measuring the pulses of high-frequency current discharges created during the process [1, 19, 22–26].

There are various methods for the measurement of partial discharge within the cell insulation. Some methods are based on a capacitive link between the whole of the winding and the measurement equipment. These setups allow the measurement of PD activity in whole windings or one phase at a time. To measure the PD activity of smaller sections of the winding, the section under test has to be disconnected from the rest of the winding.

Methods that allow the measure of PD activity in individual coils connected to the winding are based on an electromagnetic probe or pickup, which is located at the other end of a hand-held electrically insulated stick.

PD tests are conducted with voltages from below the inception voltage up to rated voltage. This fact, along with related safety reasons, practically relegates hand-held probe tests to large-diameter vertical machines (hydrogenerators). Because it is able to detect PD in individual coils, the test can give an indication of the relative deterioration of the line coils and coils in their vicinity, compared with the coils in the vicinity of the neutral. The results can be used to decide when to swap connections between the line and neutral terminations, increasing the expected life of the machine's insulation.

The signals originating from partial discharge occurring inside the cell tend to be masked by the signals originating from the more powerful slot discharge (SD). Fortunately, PD activity and SD activity result in separated frequencies. PD activity generates signals with frequency in the MHz region, while SD activity originates signals with frequencies of several kHz. By tuning the pickup circuits around these two ranges of frequencies, discrimination between the signals can be achieved.

On-line partial discharge analysis can be performed by modern instrumentation. The instruments include software that continuously analyzes the signals received from potential transformers installed on the lines and neutral of the generator. When certain thresholds are reached, the device typically sets an alarm "on."

8.1.12 Slot Discharge (SD) Test

The SD test is designed to measure partial discharge activity between coil and slot [1, 19, 27]. The existing tests are of the probe type or of the whole winding type. Slot discharges can be picked up easily with radio frequency (RF) monitors tuned to the low-frequency radio-wave spectrum. The probes will also pick corona activity at the end-windings and leads, but at a higher frequency.

8.1.13 Stator Interlaminar Insulation Tests

These tests are designed to evaluate the integrity of the insulation between the laminations of the core. They are normally carried out by inducing in the core flux densities of a magnitude similar to those that occur during normal operation of the machine. The flux is produced by looping a cable around the core and circulating a current of rated frequency. The number of ampere-turns required can be calculated from any of the many references available in the literature. IEEE Std. 432-1992 [28] gives the following expression:

$$\text{Voltage per turn of test coil} = \frac{1.05 \bullet V}{2 \bullet K_d \bullet K_p \bullet N}$$

where:

- V = line-to-line voltage for delta connection, and $\sqrt{3} \times$ line-to-line voltage for wye connection;
- Kd = distribution factor = 0.955 for a 3-phase wye-connected stator winding;
- KP = chord factor of the stator winding; and
- N = number of turns/phase in series of the stator winding.

Once the flux is established in the core, it is kept for at least 30 minutes to 1 hour. The temperature of the core should be maintained within values not significantly higher than those encountered during operation. Under these conditions, the existence of hot spots is investigated, either by placing thermometers (or other temperature transducers) in several locations, or more accurately, with infrared vision equipment. A difference of 10° to 15°C or higher between adjacent parts of the core tends to indicate a hot spot.

An additional test, called E1-CID, uses a probe that is moved along the slot and requires 3% to 5% of rated flux. The test can recognize areas of damaged iron by comparing probe readings in good and bad iron areas.

8.1.14 Core-Compression Bolts Insulation Test

The insulation between the core-compression bolts and the iron keeps the bolts from short-circuiting the insulation between the laminations. Otherwise, large eddy currents generated within the core produce heat and temperatures that further damage the interlaminar insulation as well as the insulation of the windings. This insulation can be tested with standard megger devices.

8.1.15 Bearing Insulation Test

The integrity of bearing insulation can be assessed with the use of standard megger devices. Integrity of the bearing insulation, whether of the pedestal or the bracket type, is essential to the achievement of normal life expectancy of the bearings.

8.2 MECHANICAL TESTS

Mechanical tests of the generator can be grouped as retaining ring tests, bearing tests, pressure tests, and mechanical integrity tests on the rest of the equipment, in particular the rotating members.

The retaining rings are the members subjected to what is by far the greatest scrutiny during major overhauls. The many existing designs and configurations preclude in-depth discussion of the subject in this book. One of the most comprehensive studies on this topic is given in Reference [29]. Typical tests include hardness tests and nondestructive tests or examinations such as magnetic NDE, eddy current NDE, X-rays, ultrasonic tests, and penetrating-dye NDE.

Bearing tests consist mainly of visual verification of the condition of the babbitt, as well as measurement of clearances, including air/gas seals, and so forth.

Pressure tests are conducted on liquid-cooled heat exchangers, rotor (verification of hydrogen seal integrity, performed mainly if the rotor is reworked), stator water tubing in water-cooled stators, and rotor tubing, in water-cooled rotors.

In general, the integrity of major rotating components (such as fan hubs and blades) should be subject to nondestructive examinations during major overhauls unless there is good reason to believe they are in excellent condition. Other suspected parts of the machine (such as compression plates and parts of the frame in old cast-iron machines) should also be subject to nondestructive examinations.

REFERENCES

[1] "Rotating Machinery Insulation—Test Guide," Sec. 5, Doble Engineering Company, MA, 1975–1985.

[2] IEEE Std 115-1983, "Test Procedures for Synchronous Machines," Sec. 2.10.

[3] IEEE Std 56-1977, "Guide for the Insulation Maintenance of Large Alternating-Current Rotating Machinery (10 000 kVA and Larger)," Sec. 8.1.6.

[4] IEEE Std 115-1983, "Test Procedures for Synchronous Machines," Sec. 2.01.

[5] ANSI/IEEE Std 43-1974, "Recommended Practice for Testing Insulation Resistance of Rotating Machinery."

[6] The Epoxilite Corporation, "I & C Notes," 4th article of the series *Insulation Maintenance.*

[7] The *Synchronizer E&M,* issue 200-syn-55, pp. 15–16.

[8] IEEE Std 62-1958, "Making Dielectric Measurements in the Field."

[9] ANSI/IEEE Std 433-1974/79, "Recommended Practice for Insulation Testing of Large AC Rotating Machinery with High Voltage at Very Low Frequency."

[10] IEEE Std 95-1977, "Recommended Practice for Insulation Testing of Large AC Rotating Machinery with High Direct Voltage."

[11] AIEE No. 56, "Insulation Maintenance Guide for Large Rotating Machinery."

[12] B. V. Bhimani, "Very Low Frequency High Potential Testing," IEEE TP 61-138.

[13] B. V. Bhimani, "Resistance and Capacitance Measurement on High-Voltage Insulation at Very Low Frequencies," IEEE TP 61-139.

[14] L. G. Virsberg and A. Kelen, "Some Observations on the Very Low Frequency Testing of High-Voltage Machine Insulation," *Proc. of the Conference Internationale des Grands Reseaux Electriques,* Vol. 108, June 1–10, 1964.

[15] ANSI C50.103-1977, "ANSI General Requirements for Synchronous Machines."

[16] IEEE Std 522-1977, "Guide for Testing Turn-to-Turn Insulation on Form-Wound Stator Coils for Alternating-Current Rotating Electric Machines—*For Trial Use.*"

[17] J. A. Oliver, H. H. Woodson, and J. S. Johnson, "A Turn Insulation Test for Stator Coils," *Trans. of the IEEE (PAS),* Vol. 87, 1968, pp. 669–678.

[18] J. E. Housley, "Turn-to-Turn Failure of Generator Field-Coil Insulation," Conference of Doble Clients on Rotating Machinery, Vol. 7, p. 101, 1957.

[19] *EPRI Power Plant Electrical Reference Series,* Vol. 16, "Handbook to Assess the Insulation Condition of Large Rotating Machines," 1989.

[20] M. A. El-Sharkawy, R. J. Marks, S. Oh, S. J. Huang, I. Kerszenbaum, and A. Rodriguez, "Localization of Winding Shorts Using Fuzzified Neural Networks," *Proc. of the 1994 Winter Conference of the IEEE/PES.*

[21] IEEE Std 286-1975, "Recommended Practice for Measurement of Power-Factor Tip-Up of Rotating Machinery Stator Coil Insulation."

[22] T. W. Dakin, "The Relation of Capacitance Increase with High Voltages to Internal Discharges and Discharging Void Volume," *AIEE Power Apparatus and Systems,* Vol. 78, pp. 790–795, 1959.

[23] T. W. Dakin, "A Capacitance Bridge Method for Measuring Integrated Corona-Charge Transfer and Power Loss per Cycle," *AIEE Power Apparatus and Systems,* Vol. 79, pp. 648–653, 1960.

[24] T. Orbeck, "The Measurement of Partial Discharges in High-Voltage Coils with a Capacitance Bridge," *ASEA Journal,* Vol. 40, 1967, No. 9, pp. 125–128.

[25] T. W. Dakin, C. N. Works, and J. S. Johnson, "An Electromagnetic Probe for Detecting and Locating Discharges in Large Rotating Machine Stators," *IEEE Power Apparatus and Systems,* Vol. 88, 1969, No. 3, pp. 251–257.

[26] W. McDermid, "Review of the Application of the Electromagnetic Probe Method for the Detection of Partial Discharge Activity in Stator Windings," *Proc. of the CEA International Symposium on Generator Insulation Tests,* Toronto, June 1980.

[27] J. S. Johnson, "Slot Discharge Detection between Coil Surface and Core of High-Voltage Stator Windings," *AIEE Transactions,* Vol. 70, 1951, pp. 1993–1997.

[28] IEEE Std 432, "Insulation Maintenance Guide for Medium and Small Rotating Machinery," 1992.

[29] EPRI Special Report EL/EM-5117-SR, "Guidelines for Evaluation of Generator Retaining Rings," April 1987.

Appendix

Principles of Operation of Synchronous Machines

A.1 GENERAL DISCUSSION

The commercial birth of the alternator (synchronous generator) can be dated back to August 24, 1891. On that day, the first large-scale demonstration of transmission of AC power was carried out. The transmission was from Lauffen, Germany, to Frankfurt, about 110 miles away. The occasion was an international electrical exhibition in Frankfurt. This demonstration was so convincing as to the feasibility of the utilization of AC systems for transmission of power over long distances that the city of Frankfurt adopted it for their first power plant, commissioned in 1894, exactly one hundred years before the writing of this book (see Fig. A-1).

The Lauffen–Frankfurt demonstration, and the consequent decision by the city of Frankfurt to use alternating power delivery, were instrumental in the adoption by New York's Niagara Falls power plant of the same technology. The Niagara Falls power plant became operational in 1895. For all practical purposes, the great DC versus AC duel was over. Southern California Edison's history book reports its Mill Creek hydro plant is the oldest active polyphase (3-phase) plant in the United States. Located in San Bernardino County, California, its first units went into operation on September 7, 1893, placing it almost two years ahead of the Niagara Falls project. One of those earlier units is still preserved and displayed at the plant.

It is interesting to note that, although tremendous development in machine ratings, insulation components, and design procedures has occurred over the last 100 years, the basic constituents of the machine have remained the same.

Fig. A-1 The hydroelectric generator from Lauffen, now in the Deutsches Museum, Munich. (Extracted with permission from "The Evolution of the Synchronous Machine" by Gerhard Neidhofer, 1992, ABB.)

The concept that a synchronous generator can be used as a motor followed suit. Although Tesla's induction motor replaced the synchronous motor as the choice for the vast majority of electric motor applications, synchronous generators remained the universal machine for the generation of electric power. The world today is divided between countries generating their power at 50 Hz and others (like the United States) at 60 Hz. Additional frequencies (like 25 Hz) can still be found in some locations, but they constitute the rare exception.

Synchronous generators have continuously grown in size over the years. The justification is based on simple economies of scale: The output rating of the machine per unit of weight increases as the size of the unit increases. Thus, it is common to see machines with ratings up to 1500 MVA, normally used in nuclear power stations. Interestingly enough, the present ongoing shift from large steam turbines as prime movers to more efficient gas turbines is resulting in a reverse of the trend toward larger and larger generators, at least for the time being. Transmission system stability considerations also place an upper limit on the rating of a single generator.

A.2 CONSTRUCTION

Synchronous machines come in all sizes and shapes, from the permanent magnet synchronous motor in wall clocks, to the largest steam-turbine-driven generators of 1500 MW or more. Synchronous machines are one of two types: the stationary field or the rotating DC magnetic field.

The *stationary field* synchronous machine has salient poles mounted on the stator—the stationary member. The poles are magnetized either by permanent magnets or by a DC current. The armature, normally containing a 3-phase winding, is mounted on the shaft. The armature winding is fed through three slip-rings (collectors) and a set of brushes sliding on them. This arrangement can be found in machines up to about 5 kVA in rating. For larger machines—all those covered in this book—the typical arrangement used is the rotating magnetic field.

The *rotating magnetic field* (also known as *revolving-field*) synchronous machine has the field winding wound on the rotating member (the *rotor*), and the armature wound on the stationary member (the *stator*). The rotating winding is energized by a DC current, creating a magnetic field that must be rotated at synchronous speed. The rotating field winding can be energized through a set of slip-rings and brushes (external excitation), or from a diode bridge mounted on the rotor (self-excited). The rectifier bridge is fed from a shaft-mounted alternator, which is itself excited by the pilot exciter. In externally fed fields, the source can be a shaft-driven DC generator, a separately excited DC generator, or a solid-state rectifier. Several variations to these arrangements exist.

The stator core is made of insulated steel laminations. The thickness of the laminations and the type of steel are chosen to minimize eddy current and hysteresis losses. The core is mounted directly onto the frame or (in large 2-pole machines) through spring bars. The core is slotted (normally open slots), and the coils making the winding are placed in the slots. There are several types of armature windings; e.g., concentric windings of several types, cranked coils, split windings of various types, wave windings, and lap windings of various types. Modern large machines typically are wound with double-layer lap windings.

The rotor field is either of *salient-pole* (Fig. A-2a) or *non-salient-pole* construction, also known as *round rotor* or *cylindrical rotor* (Fig. A-2b).

Non-salient-pole rotors are utilized in 2- or 4-pole machines, and occasionally in 6-pole machines. These are typically driven by steam- or gas-turbine prime movers. The vast majority of salient-pole machines have six or more poles. They include all synchronous hydrogenerators, almost all synchronous condensers, and the overwhelming majority of synchronous motors.

Non-salient-pole rotors are typically machined out of a solid steel forging. The winding is placed in slots machined out of the rotor body and retained against the large centrifugal forces by metallic wedges, normally made of aluminum or steel. The end part of the windings is retained by the so-called retaining rings. In the case of large machines, these are made out of steel.

Large salient-pole rotors are made of laminated poles retaining the winding under the pole head. The poles are keyed onto the shaft or spider-and-wheel structure. Salient-pole machines have an additional winding in the rotating member. This winding, made of copper bars short-circuited at both ends, is imbedded in the head of the pole, close to the face of the pole. The purpose of this winding is

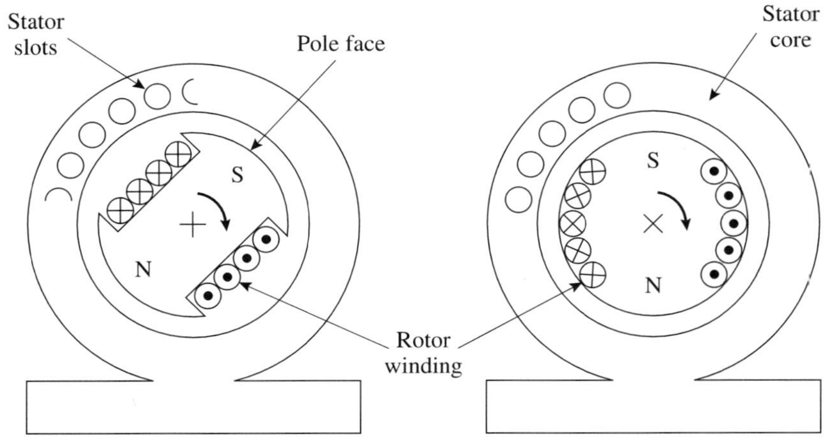

Fig. A-2 Schematic cross section of a synchronous machine. (a) Salient pole; (b) Round rotor.

to start the motor or condenser under its own power as an induction motor, and take it unloaded to almost synchronous speed, when the rotor is "pulled in" by the synchronous torque. The winding also serves to damp the oscillations of the rotor around the synchronous speed, and is therefore named the *damping winding* (also known as *amortisseurs*).

A.3 OPERATION

It is convenient to introduce the fundamental principles describing the operation of a synchronous machine in terms of an *ideal* cylindrical-rotor machine connected to an *infinite* bus. The infinite bus represents a busbar of constant voltage, which can deliver or absorb active and reactive power without any limitations. The ideal machine has zero resistance and leakage reactance, infinite permeability, and no saturation, as well as zero reluctance torque.

The production of torque in the synchronous machine results from the natural tendency of two magnetic fields to align themselves. The magnetic field produced by the stationary armature is denoted as ϕ_s. The magnetic field produced by the rotating field is ϕ_f. The resultant magnetic field is

$$\phi_r = \phi_s + \phi_f$$

The flux ϕ_r is established in the airgap of the machine. *(Bold symbols indicate vector quantities.)*

When the torque applied to the shaft is zero, the magnetic fields of the rotor and stator completely align themselves. The instant torque is introduced to the shaft, either in a generating mode or in a motoring mode, and a small angle is created between the stator and rotor fields. This angle (λ) is called the *torque angle* of the machine.

A.3.1 No-Load Operation

When the ideal machine is connected to the infinite bus, a 3-phase balanced voltage ($\mathbf{V}_1$) is applied to the stator winding (within the context of this work, 3-phase systems and machines are assumed). It can be shown that a 3-phase balanced voltage applied to a 3-phase winding evenly distributed around the core of an armature will produce a rotating (revolving) magnetomotive force (mmf) of constant magnitude ($\mathbf{F}_s$). This mmf, acting upon the reluctance encountered along its path, results in the magnetic flux (ϕ_s) previously introduced. The speed at which this field revolves around the center of the machine is related to the supply frequency and the number of poles, by the following expression:

$$n_s = 120 \bullet \frac{f}{p}$$

where (f) = electrical frequency in Hz
(p) = number of poles of the machine
(n_s) = speed of the revolving field in revolutions per minute (rpm).

If no current is supplied to the DC field winding, no torque is generated, and the resultant flux (ϕ_r), which in this case equals the stator flux (ϕ_s), magnetizes the core to the extent the applied voltage ($\mathbf{V}_1$) is exactly opposed by a counter-electromotive force (cemf) ($\mathbf{E}_1$). If the rotor's excitation is slightly increased, and no torque applied to the shaft, the rotor provides some of the excitation required to produce ($\mathbf{E}_l$), causing an equivalent reduction of (ϕ_s). This situation represents the underexcited condition shown in condition *no-load* (a) in Figure A-3. When operating under this condition, the machine is said to behave as a *lagging condenser;* i.e., it absorbs reactive power from the network. If the field excitation is increased over the value required to produce ($\mathbf{E}_l$), the stator currents generate a flux that counteracts the field-generated flux. Under this condition, the machine is said to be overexcited, shown as condition *no-load* (b) in Figure A-3. The machine is behaving as a *leading condenser;* i.e., it is delivering reactive power to the network.

Under no-load condition, both the torque angle (λ) and the load angle (δ) are zero. The *load angle* is defined as the angle between the rotor's mmf ($\mathbf{F}_f$) or flux (ϕ_f) and the resultant mmf ($\mathbf{F}_r$) or flux (ϕ_r). The load angle (δ) is the most commonly used because it establishes the torque limits the machine can attain in a simple manner (discussed later). One must be aware that, in many texts, the name

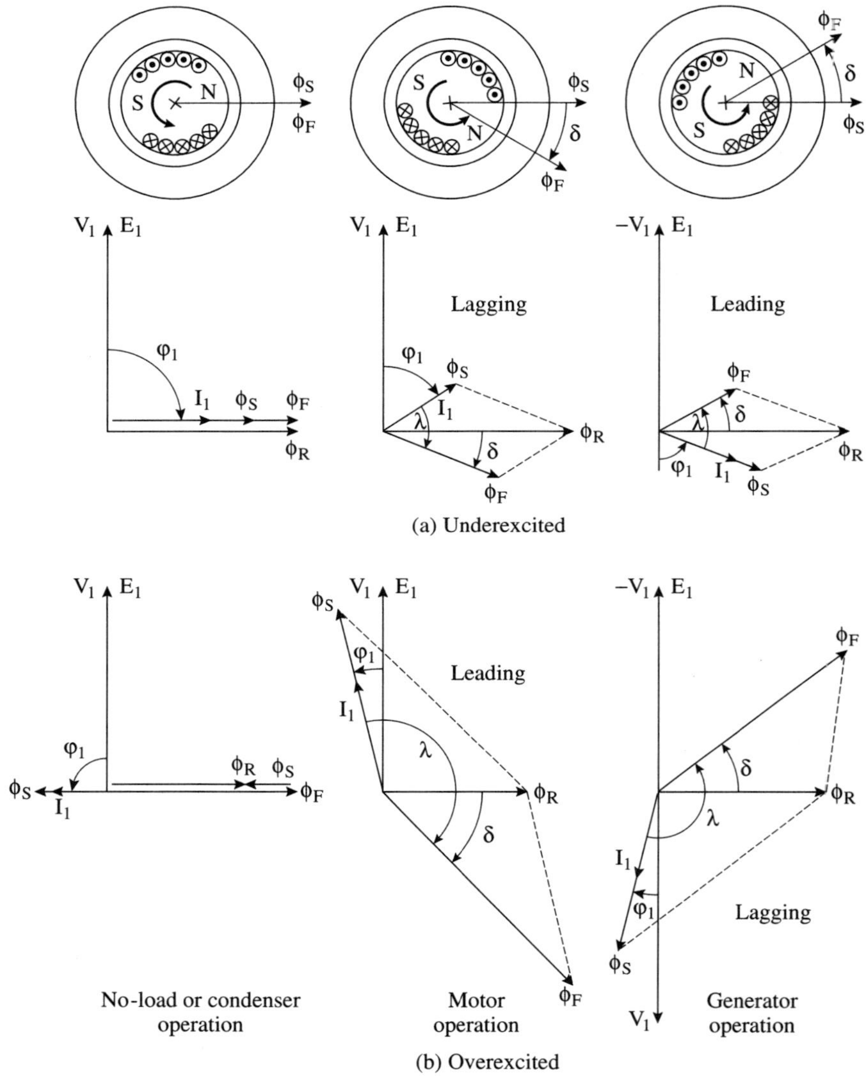

Fig. A-3 Phasor diagrams for a synchronous cylindrical-rotor ideal machine.

torque angle is used to indicate the load angle. The name *torque angle* is also sometimes given to indicate the angle between the terminal voltage ($\mathbf{V}_1$) and the excitation voltage ($\mathbf{E}_1$). This happens because the leakage reactance is generally very much smaller than the magnetizing reactance, and therefore the load angle

(δ) and the angle between ($\mathbf{V}_l$) and ($\mathbf{E}_l$) are very similar. In this book, the more common name *power angle* is used for the angle between ($\mathbf{V}_l$) and ($\mathbf{E}_l$). In Figure A-3, the power angle is always shown as zero because the leakage impedance has been neglected in the ideal machine.

It is important at this stage to introduce the distinction between electrical and mechanical angles. In studying the performance of the synchronous machine, all the electromagnetic calculations are carried out based on electric quantities; i.e., all angles are electrical angles. To convert the electrical angles used in the calculations to the physical mechanical angles we observe, the following relationship applies:

$$\text{mechanical angle} = \left(\frac{2}{p}\right) \text{electrical angle}$$

A.3.2 Motor Operation

If a breaking torque is applied to the shaft, the rotor starts falling behind the revolving-armature-induced mmf ($\mathbf{F}_s$). In order to maintain the required magnetizing mmf ($\mathbf{F}_r$), the armature current changes. If the machine is in the underexcited mode, the condition *motor* (a) in Figure A-3 represents the new phasor diagram. On the other hand, if the machine is overexcited, the new phasor diagram is represented by *motor* (b) in Figure A-3. The active power consumed from the network under these conditions is given by:

$$\text{Active power} = V_1 \bullet I_1 \bullet \cos \varphi_1 \qquad \text{(per phase)}$$

If the torque is increased, a limit is reached in which the rotor cannot keep up with the revolving field. The machine then stalls. This is known as "falling out of step," or "pulling out of step," or "slipping poles." The maximum torque limit is reached when the angle δ equals $\pi/2$ electrical. The convention is to define δ as negative for motor operation and positive for generator operation. The torque is also a function of the magnitude of ϕ_r and ϕ_f. When overexcited, the value of ϕ_f is larger than in the underexcited condition. Therefore, synchronous motors are capable of greater mechanical output when overexcited. Likewise, underexcited operation is more prone to result in an "out-of-step" situation.

A.3.3 Generator Operation

Let's assume that the machine is running at no-load and a positive torque is applied to the shaft; i.e., the rotor flux angle is advanced ahead of the stator flux angle. As in the case of motor operation, the stator currents will change to create the new conditions of equilibrium shown in Figure A-3, under *generator.* If the machine is initially underexcited, condition (a) in Figure A-3 results. On the other hand, if the machine is overexcited, condition (b) in Figure A-3 results.

It is important to note that when seen from the terminals, with the machine operating in underexcited mode, the power factor angle (φ_1) is leading (i.e., $\mathbf{I}_1$ leads $\mathbf{V}_1$). This means the machine is absorbing reactive power from the system. The opposite occurs when the machine is in overexcited mode. As for the motor operation, an overexcited condition in the generating mode also allows for greater power deliveries.

As generators are normally called to provide VARs together with watts, they are almost always operated in the overexcited condition.

A.3.4 Equivalent Circuit

The most convenient way to determine the performance characteristics of synchronous machines is by means of equivalent circuits. These equivalent circuits can become very elaborate when saturation, armature reaction, harmonic reactance, and other nonlinear effects are introduced. However, the simplified circuit in Figure A-4 is conducive to obtaining the basic performance characteristics of the machine under steady-state conditions.

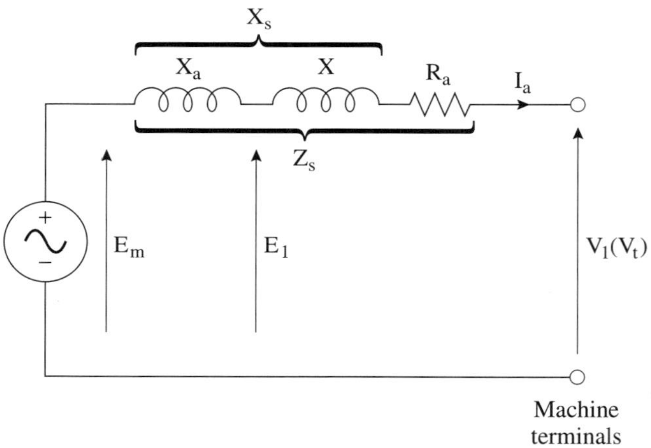

Fig. A-4 Steady-state equivalent circuit of a synchronous machine.

X = leakage reactance
X_a = armature reaction reactance
$X_s = X_a + X$ = synchronous reactance
R_a = armature resistance
Z_s = synchronous impedance
$V_l(V_t)$ = terminals voltage
E_m = magnetizing voltage

In Figure A-4, the reactance X_a represents the magnetizing or demagnetizing effect of the stator windings on the rotor. It is also called the *magnetizing reactance.* R_a represents the effective resistance of the stator. The reactance X represents the stator leakage reactance. The sum of X_a and X is used to represent the total reactance of the machine, and is called the *synchronous reactance* (X_s). Z_s is the *synchronous impedance* of the machine. It is important to remember that the equivalent circuit described in Figure A-4 represents the machine only under steady-state condition.

The simple equivalent circuit of Figure A-5a (see p. 162) suffices to determine the steady-state performance parameters of the synchronous machine connected to a power grid. These parameters include voltages, currents, power factor, and load angle (see Fig. A-5b). The regulation of the machine can be easily found from the equivalent circuit for different load conditions by using the regulation formula:

$$\Re(\%) = 100 \bullet \left(V_{\text{no-load}} - V_{\text{load}}\right) / V_{\text{load}}$$

A.3.5 Performance Characteristics: *V*-Curves and Rating Curves

If the active power flow of the synchronous machine is kept constant, a family of curves can be obtained relating the magnitude of the armature current to that of the field current. The curves, shaped as *V* (see Fig. A-6), are drawn for various load conditions. In the graph, the lagging and leading operating regions can be discerned.

Physical considerations define the limits of operation of synchronous machines. These limits are expressed as a family of concentric capability curves (see Fig. A-7).

- The top part of the rating curves is defined by the field winding heating and insulation system.
- The right side of the curves is limited by the heating of the armature and the type of armature insulation.
- The bottom part of the curves is limited by the heating of the core-end region.

Rating curves are normally drawn for a number of hydrogen pressures (in hydrogen-cooled machines), or for ambient temperatures (in air-cooled machines).

Both the rating curves and the *V*-curves can be combined in one graph. This graph is used by the machine operators and protection engineers to set the limits of safe operation on the machine.

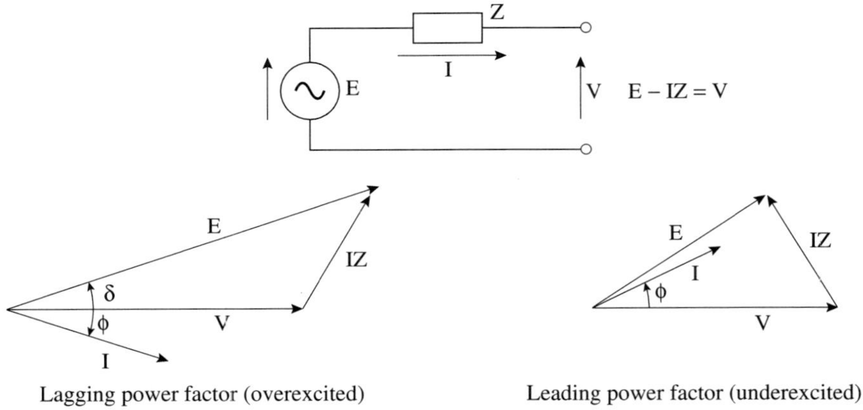

(a) Generator operation

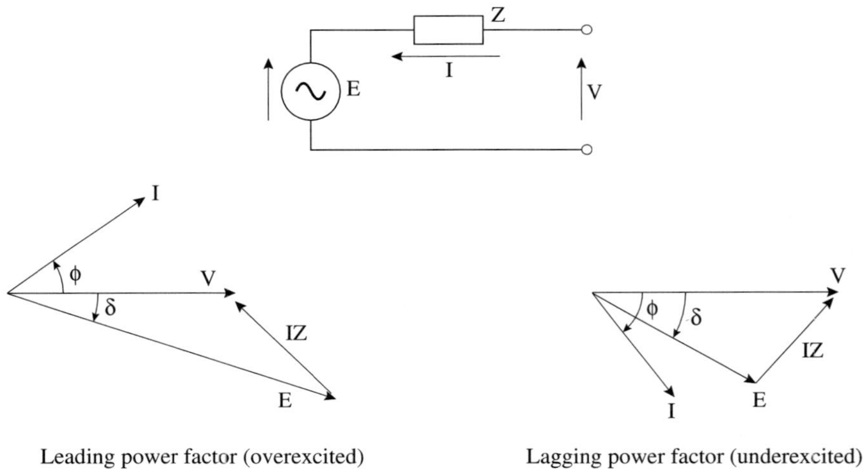

(b) Motor operation

Fig. A-5 Steady-state equivalent circuit and vector diagram.

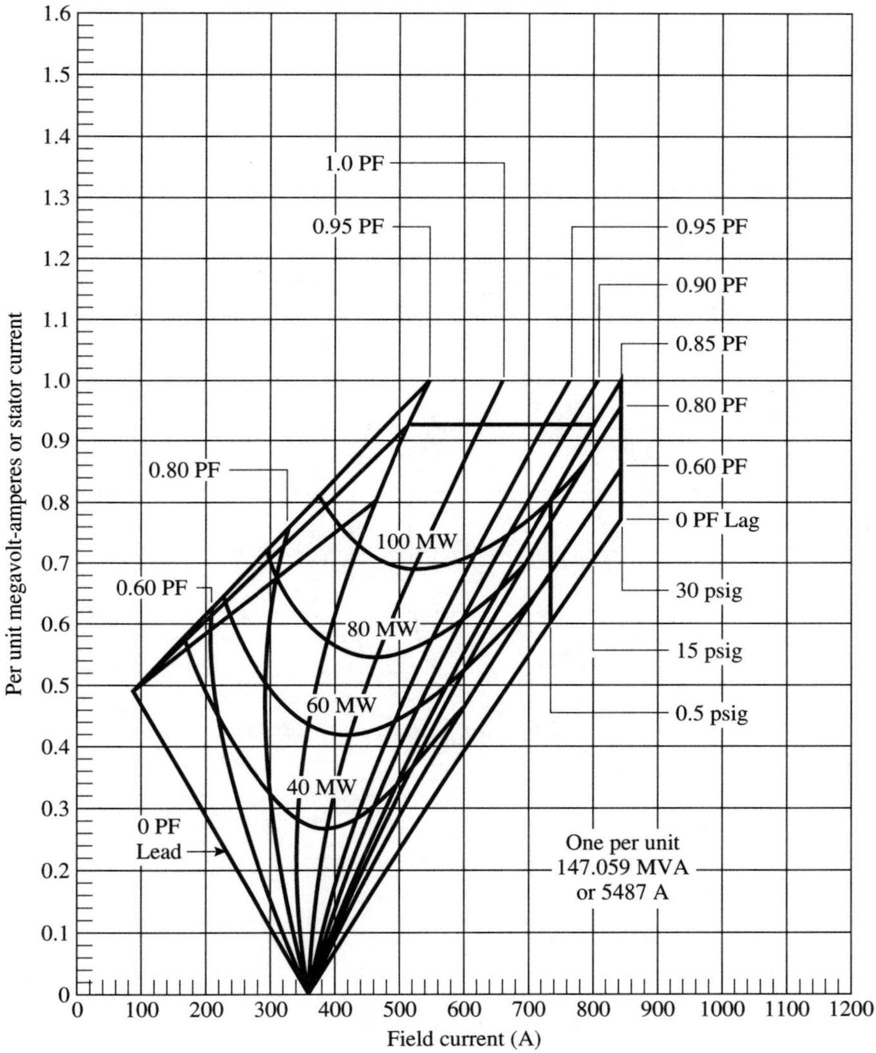

Fig. A-6 Typical *V*-curves for generator operation. (Copyright © 1987. Electric Power Research Institute.)

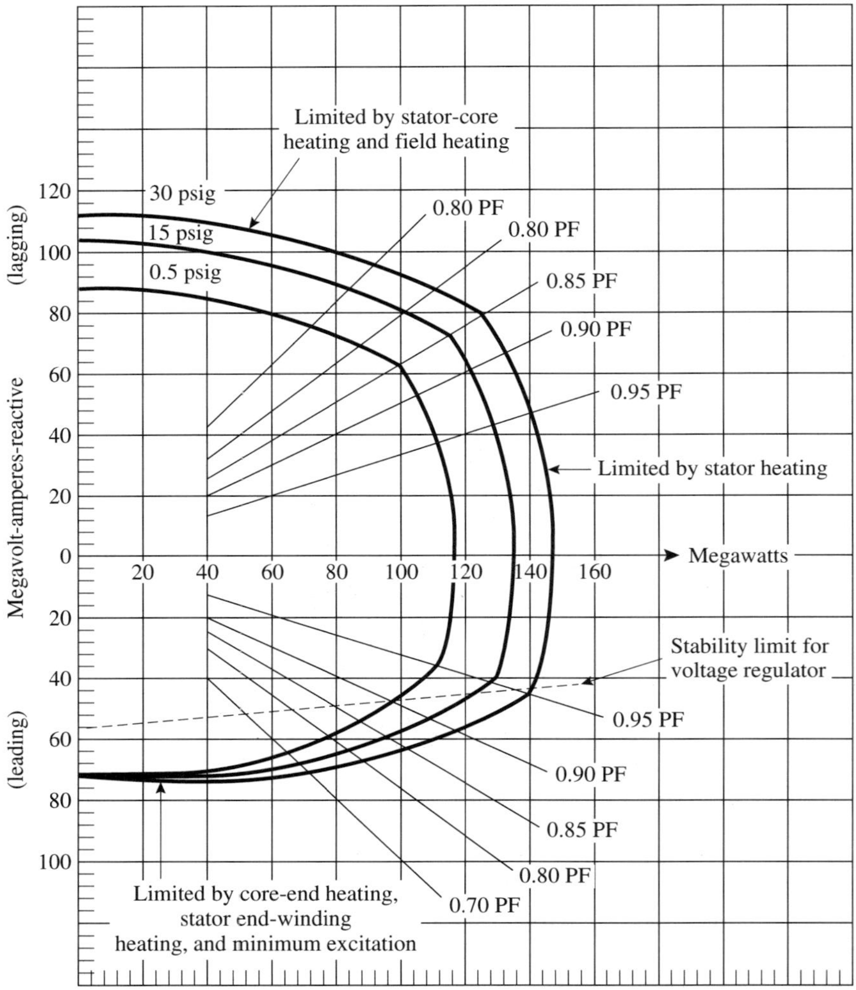

Fig. A-7 Typical capability curves for a synchronous generator. (Copyright © 1987. Electric Power Research Institute. EPRI EL-5036. *Power Plant Electrical Reference Series,* Volumes 1–16. Reprinted with Permission.)

A.4 OPERATING CONSTRAINTS

In addition to the rating curves described in Section A.3.5, design characteristics of the machine impose additional limits to its range of allowed operation. The items described in the next few sections represent some of the most important constraints imposed on the machine. ANSI and IEEE standards in the United States and other standards abroad provide in many cases typical ranges for those values. Also, typical values can be found in technical papers, books, and bulletins published by the machine manufacturers.

It is interesting to note that in certain cases (such as maximum-allowed over- and under-frequency operation of turbine-driven generators), the prime mover (steam- or gas-turbine) may impose stricter limitations on the operation of the unit than the generator.

Reference [1] is an excellent source of information on the operational requirements of large synchronous machines.

A.4.1 Volts per Hertz (V/Hz)

The term "volts per hertz" has been borrowed from the operation of transformers. In transformers, the *fundamental voltage equation* is given by:

$$V = 4.44 \bullet f \bullet \boldsymbol{B}_{\max} \bullet \text{area of core} \bullet \text{number of turns}$$

where $\boldsymbol{B}_{\max}$ is the vector magnitude of the flux density in the core of the transformer.

By rearranging the variables, the following expression is obtained:

$$V/f\ \left[V/Hz\right] = 4.44 \bullet B_{\max} \bullet \text{area of core} \bullet \text{number of turns}$$

or alternatively,

$$B_{\max}\ [\text{Tesla}] = \text{constant} \bullet (V/f)$$

or, in another annotation,

$$B_{\max} \propto \text{V}/\text{Hz}$$

This last equation indicates that the maximum flux density in the core of a transformer is proportional to the terminal voltage divided by the frequency of the supply voltage. This ratio is known as *V/Hz*.

A set of equations very similar to the ones above can be written for the armature of an alternate-current machine. In this case, the constant includes winding parameters such as winding *pitch* and *distribution* factors. However, the end result is the same; i.e., in the armature of an electrical alternate current machine, the maximum core flux density is proportional to the terminal voltage divided by the supply frequency (or V/Hz).

The importance of this ratio resides in the fact that in machines, as well as in transformers, the operating point of the voltage is such that, for the given rated frequency, the flux density is just below the knee of the saturation point.

Increasing the volts per turn in the machine (or transformer) raises the flux density above the knee of the saturation curve (see Fig. A-8). Consequently, large magnetization currents are produced, as well as large increases in the core loss due to the bigger hysteresis loop created (see Fig. A-9). Both of these result in substantial increases in core and copper losses and excessive temperature rises in both core and windings. If not controlled, this condition can result in loss of the core interlaminar insulation, as well as loss of life of the winding insulation.

The ANSI C50.30-1972/IEEE Std 67-1972 standard state generators are normally designed to operate at rated output of up to 105% of rated voltage [1]. ANSI/IEEE C57 standards for transformers state the same percentage for rated loads and up to 110% of rated voltage at no-load. In practice, the operator should make sure the machine remains below limits that may affect the integrity of both the generator and the unit transformer. The aforementioned standards state that synchronous motors, like motors in general, are typically designed for rated operation under voltages of up to 110% of rated voltage. For operation of synchronous machines at other than rated frequencies, refer to ANSI C50.30-1972 [1].

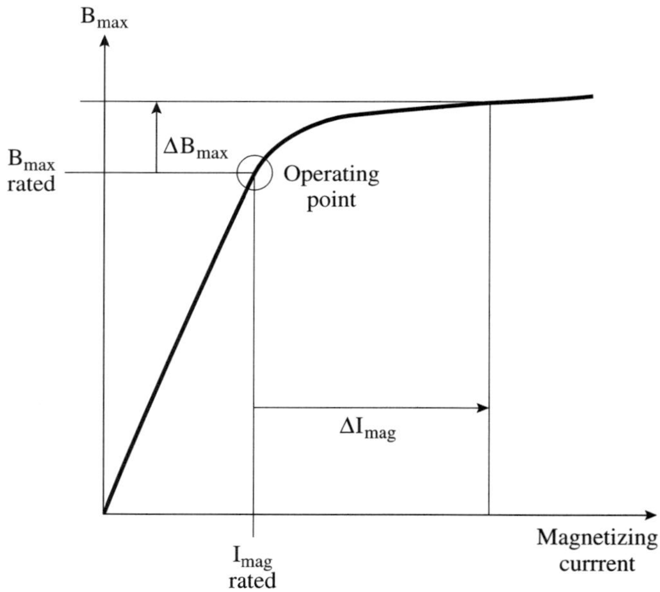

Fig. A-8 Typical saturation curve for transformers and machines.

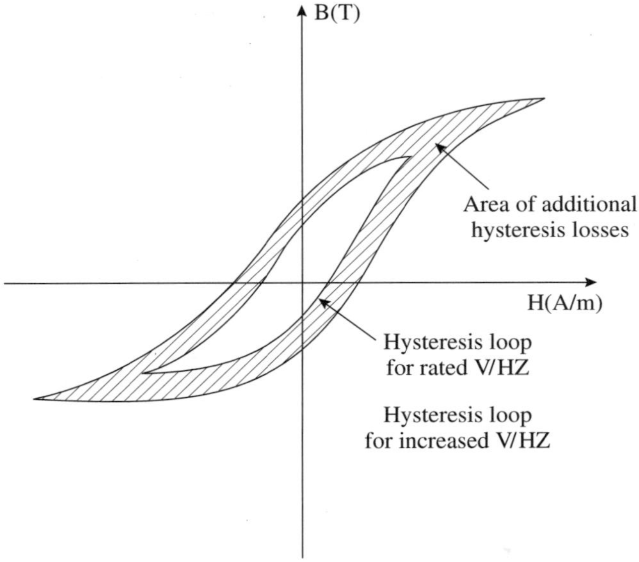

Fig. A-9 Hysteresis losses under normal and abnormal conditions.

A.4.2 Negative Sequence Currents and $(I_2)^2t$

A 3-phase balanced supply voltage applied to a symmetrical 3-phase winding generates a constant-magnitude flux in the airgap of the machine, which rotates at synchronous speed around the circumference of the machine. In addition, the slots and other asymmetries within the magnetic path of the flux create low-magnitude space harmonics; i.e., fluxes that rotate in both directions, of multiple frequencies of the fundamental supply frequency. In a synchronous machine, the main (fundamental) flux rotates in the same direction and speed as the rotor.

It happens that when the supply voltage or currents are unbalanced, an additional flux of fundamental frequency appears in the airgap of the machine. However, this flux rotates in the opposite direction from the rotor. This flux induces in the rotor windings and body voltages and currents with twice the fundamental frequency. These are called *negative-sequence currents and voltages.*

There are several abnormal operating conditions that give rise to large currents flowing in the forging of the rotor, rotor wedges, teeth, end rings, and field windings of synchronous machines. These conditions include unbalanced armature current (producing negative-sequence currents) as well as asynchronous motoring or generation (operation with loss of field), producing alternate stray rotor currents. The resultant stray rotor currents tend to flow on the surface of the rotor, generating $(I_2)^2$ R losses with rapid overheating of critical rotor components. If not properly controlled, serious damage to the rotor will ensue. Of particular concern is damage to the end rings and wedges of round rotors (see Fig. A-10).

For all practical purposes, all large synchronous machines have installed protective relays that will remove the machine from operation under excessive negative sequence current operation. To properly "set" the protective relays, the operator should obtain maximum allowable negative sequence "I_2" values from the machine's manufacturer. The values shown in Table A-1 are contained in ANSI/IEEE C50.13-1977 [2] as values of continuous I_2 current to be withstood by a generator without injury, while exceeding neither rated kVA nor 105% of rated voltage.

TABLE A-1. Values of Permissible I_2 Current in a Generator

Type of Generator	Permissible I_2 as % of Rated Stator Current
Salient-pole:	
With connected amortisseur winding	10
Without connected amortisseur winding	5
Cylindrical-rotor:	
Indirectly cooled	10
Directly cooled up to 950 MVA	8
951–1,200 MVA	6
1,200–1,500 MVA	5

When unbalanced fault currents occur in the vicinity of a generator, the I_2 values of Table A-1 will probably be exceeded. In order to set the protection relays to remove the machine from the network before damage is incurred, but avoiding unnecessary relay misoperation, manufacturers have developed the so-called $(I_2)^2t$ values. These values represent the maximum time in seconds a machine can be subjected to a negative-sequence current. In the $(I_2)^2t$ expression, the current is given as per unit of rated stator current. These values should be obtained from the manufacturer. Table A-2 shows typical values given in the standard [2].

TABLE A-2. Values of Permissible $(I_2)^2t$ in a Generator

Type of Machine	Permissible $(I_2)^2t$
Salient-pole generator	40
Salient-pole condenser	30
Cylindrical-rotor generator:	
Indirectly cooled	30
Directly cooled, 0–800 MVA	30
Directly cooled, 801–1,600 MVA	(10-5/800)(MVA-800)

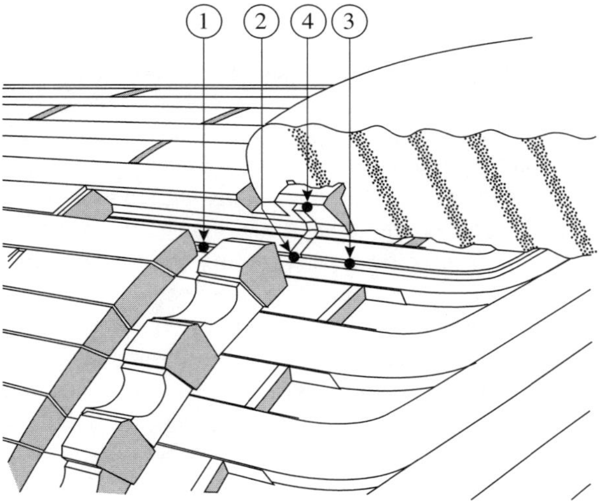

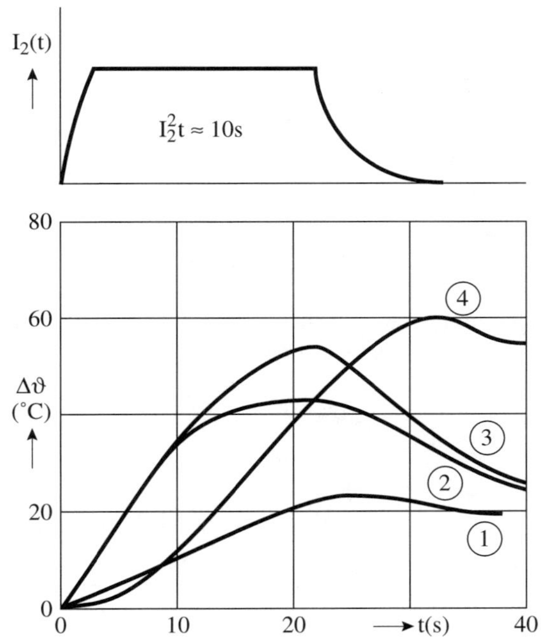

Fig. A-10 Temperature rise measured at the end of the rotor body during short-term unbalanced load operation. (Reproduced with permission from "Design and Performance of Large Steam Turbine Generators," 1974, ABB.)

A.4.3 Overspeed

Manufacturers of large rotating machines normally test their products to withstand speeds of up to 20% over rated speed. Nevertheless, aging of the machine may, to some extent, encroach on the original design margins. Therefore, overspeed is a serious condition that must be avoided by proper setting of the protective instrumentation. In steam-turbine generators, the turbine is often the item most sensitive to overspeed operation of the unit, and protection is set accordingly. Hydrogenerators tend to overspeed for longer periods during a sudden loss of load, due to the slower water valves.

Many old salient-pole hydrogenerators still in operation, which were originally designed for 50-Hz operation, were converted years ago to 60-Hz operation (a 20% speed increase), in addition to large up-rating of delivered load. The change was predicated on the large design margins of these old machines. However, in most cases it is difficult to know the remaining overspeed margin of these machines. Detailed mechanical calculations are required.

REFERENCES

[1] ANSI C50.30-1972/IEEE Std 67-1972, "IEEE Guide for Operation and Maintenance of Turbine Generators."

[2] ANSI/IEEE C50.13-1977, "Requirements for Cylindrical-Rotor Synchronous Generators."

ADDITIONAL READING

A wealth of literature exists for the reader interested in a more in-depth understanding of synchronous machine theory. The following is only a very short list of textbooks readily available describing the operation and design of synchronous machines in a manner accessible to the uninitiated.

Dino Zorbas, *Electric Machines—Principles, Applications, and Control Schematics.* West Publishing Company, 1989.

M. G. Say, *Alternating Current Machines.* Pitman Publishing Limited, 1978.

Theodore Wildi, *Electrical Machines, Drives and Power Systems.* Prentice Hall.

Vincent del Toro, *Electric Machines and Power Systems.* Prentice Hall, 1985.

M. Liwschitz-Garik and C. C. Whipple, *Electric Machinery, Vols. 1 & 2.* D. Van Nostrand Co., Inc.

A. E. Fitzgerald and C. Kingsley, *Electric Machinery.* McGraw-Hill.

Index

C

D

E

M

N

O

P

R

V

W

Z